T0340094

Practical Power System and Protective Relays Commissioning

Practical Power System and Protective Relays Commissioning

Omar Salah Elsayed Atwa
Electrical Power Engineer

ACADEMIC PRESS
An imprint of Elsevier

Academic Press is an imprint of Elsevier
125 London Wall, London EC2Y 5AS, United Kingdom
525 B Street, Suite 1650, San Diego, CA 92101, United States
50 Hampshire Street, 5th Floor, Cambridge, MA 02139, United States
The Boulevard, Langford Lane, Kidlington, Oxford OX5 1GB, United Kingdom

Notices
Knowledge and best practice in this field are constantly changing. As new research and experience broaden
our understanding, changes in research methods, professional practices, or medical treatment may become
necessary.

Practitioners and researchers must always rely on their own experience and knowledge in evaluating and
using any information, methods, compounds, or experiments described herein. In using such information or
methods they should be mindful of their own safety and the safety of others, including parties for whom they
have a professional responsibility.

To the fullest extent of the law, neither the Publisher nor the authors, contributors, or editors, assume any
liability for any injury and/or damage to persons or property as a matter of products liability, negligence or
otherwise, or from any use or operation of any methods, products, instructions, or ideas contained in the
material herein.

British Library Cataloguing-in-Publication Data
A catalogue record for this book is available from the British Library

Library of Congress Cataloging-in-Publication Data
A catalog record for this book is available from the Library of Congress

ISBN: 978-0-12-816858-5

For Information on all Academic Press publications
visit our website at https://www.elsevier.com/books-and-journals

Publisher: Joe Hayton
Acquisition Editor: Lisa Reading
Editorial Project Manager: Joanna Collett
Production Project Manager: Anitha Sivaraj
Cover Designer: Christian J. Bilbow

Typeset by MPS Limited, Chennai, India

Working together
to grow libraries in
developing countries

www.elsevier.com • www.bookaid.org

Contents

Introduction

During my working experience of 28 years from the time I started work in the protection and commissioning field I have collected the important and essential information required to do this job by reading many books and scientific papers, and attending numerous training courses. However, I still struggled to find the required information, and therefore decided to collect this important information in this field into one book as a guide for new engineers starting in this field and also for experienced professional engineers in this field to transfer their knowledge to the next generation. I hope that this book will be useful for students and engineers, and I wish to thank all those who helped me in publishing this book, including my wife who encouraged me along the project.

This book consists of 23 chapters. The first 15 chapters deal with power system component theory and testing. Chapter 16, Power System Fault Analysis and Chapter 17, IEC 61850 Protocols Used in Protective Relays Communication, and Chapter 18, Protection Relays, deals with the theory of protective relays. Chapter 19, Protection Relays Settings describe protective relay setting procedures, and Chapter 20, Protective Relays Testing and Commissioning describe protective relays testing and commissioning procedures. Chapter 21, A Guided Practical Value of Some Test Results Collected From Actual Power System Testing at Site. Chapter 22, Final Substation Primary and Energization and Loading Tests, which deals with the final tests for substations before energization, and the final chapter deals with the time scheduling and resource planning of the commissioning of substations with a practical estimation of the substation commissioning schedule and resources based on the standard commissioning tests that are required to be performed.

Chapter 1

Power System Elements

1.1 INTRODUCTION

A power system consists of a group of elements working together, including generation, transmission, distribution, and loads. Each of these are discussed in the following sections.

1.2 OVERVIEW OF A POWER SYSTEM

1.2.1 Generation of Power

Power generation converts energy from one form to another. Examples are:

1. Thermal generating stations that use fuels such as coal, oil, and gas that are burnt to generate electricity.
2. Hydro generating stations that depend on water flow through a turbine to drive the generator.
3. Nuclear generation stations that depend on uranium to generate electricity (nuclear fuel).

1.2.2 Transmission System

A transmission system is a network of overhead lines, underground cables, and transformers that transfer generated power in high voltage levels to loads.

High voltage levels are used to reduce the losses in the system during power transfer between each end of the transmission line.

1.2.3 Distribution System

A distribution system is a network of overhead transmission lines, underground cables, and transformers that carry small amounts of power to loads in medium and low voltage levels.

Practical Power System and Protective Relays Commissioning.
DOI: https://doi.org/10.1016/B978-0-12-816858-5.00001-0

1.2.4   Loads

Loads are defined as items that consume electricity, for example, homes and factories.

1.3   SYSTEM VOLTAGES

Voltages differ from country to country, however, a general classification is:

Generation voltage levels range from 12 to 33 kV.
Transmission voltage levels range from 110 to 800 kV.
Distribution voltage levels range from 4 to 35 kV.

1.4   POWER SYSTEM COMPONENTS

A power system is made up of multiple components. These are:
Generators that generate electrical power.
Transformers that change voltage levels as per system requirements.
Circuit breakers that open and close the power circuits under normal and abnormal conditions.
Shunt reactors and capacitors that are used for system compensation for reactive power in transmission lines.
Surge arresters are used for protecting the primary system from overvoltages.
Protective relays are used to detect any faults in the system and to trip the circuit breakers to isolate the faulty part in the power system to keep other parts of the system healthy and working in normal condition.
Substations are nodes of several transmission lines or distribution lines with power transformers.
Buses are heavy conductors that collect power and distribute it to the different system components.
These are all shown in the simplified power system shown in Fig. 1.1.

1.5   IEEE DEVICE NUMBERS AND FUNCTIONS FOR SWITCHGEAR APPARATUS

The devices in switching equipment are referred to by numbers, according to the functions they perform. These numbers are based on a system which has been adopted as standard for automatic switchgear by the Institute of Electrical and Electronics Engineers (IEEE). This system is used on connection diagrams, in instruction books, and in specifications. The standard numbers referring to device functions in switchgear components are as follows.
32   Directional power relay
21   Distance relay
25   Synchronizing or synchrocheck device

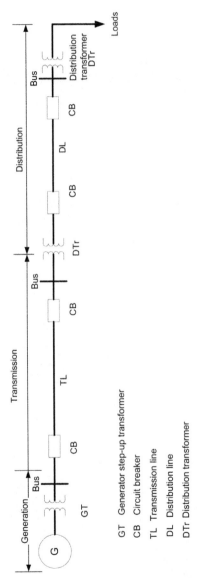

FIGURE 1.1 Simple power system.

GT Generator step-up transformer
CB Circuit breaker
TL Transmission line
DL Distribution line
DTr Distribution transformer

27 Under voltage relay

37 Under current or under power relay

40 Field relay

41 Field circuit breaker

42 Running circuit breaker

46 Reverse phase or phase−balance current relay

47 Phase−sequence voltage relay

49 Machine or transformer thermal relay

50 Instantaneous overcurrent or rate of rise relay

51 AC time overcurrent relay

52 AC circuit breakers

55 Power factor relay

59 Overvoltage relay

63 Liquid or gas pressure, level, or flow relay

65 Governing equipment which controls the gate or valve opening of a prime mover

67 AC directional overcurrent relay

72 DC circuit breakers

76 DC overcurrent relay

79 AC autoreclosing relay

81 Frequency relay

85 Carrier or pilot-wire receiver relay

86 Locking-out relay

87 Differential relay

89 Line switch

94 Tripping relay

There are symbols for switchgear and protective devices as shown in Fig. 1.2.

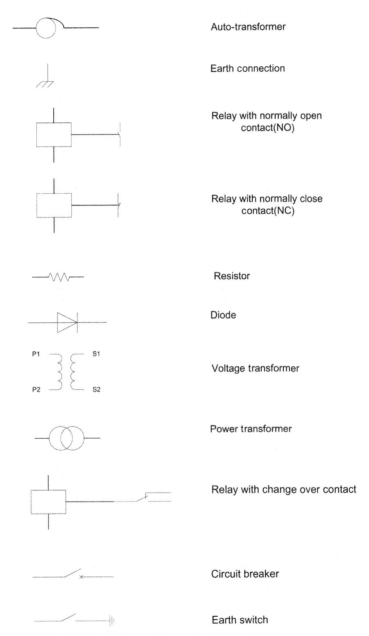

	Auto-transformer
	Earth connection
	Relay with normally open contact(NO)
	Relay with normally close contact(NC)
	Resistor
	Diode
	Voltage transformer
	Power transformer
	Relay with change over contact
	Circuit breaker
	Earth switch

FIGURE 1.2   Symbols for switchgear and protective devices.

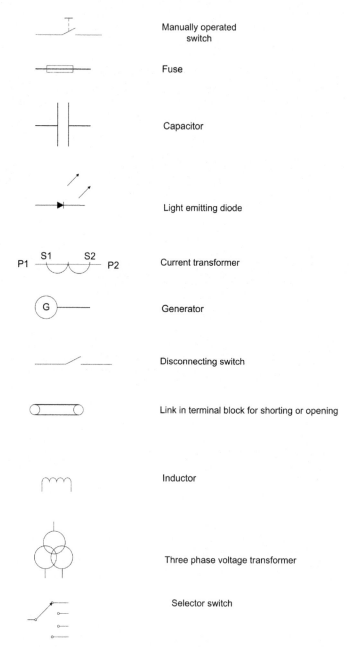

	Manually operated switch
	Fuse
	Capacitor
	Light emitting diode
P1 S1 S2 P2	Current transformer
G	Generator
	Disconnecting switch
	Link in terminal block for shorting or opening
	Inductor
	Three phase voltage transformer
	Selector switch

FIGURE 1.2   (Continued)

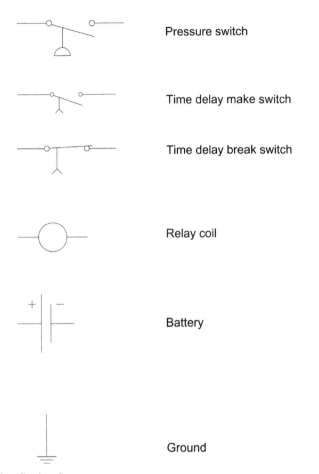

Pressure switch

Time delay make switch

Time delay break switch

Relay coil

Battery

Ground

FIGURE 1.2   (Continued)

Chapter 2

Substations

2.1 INTRODUCTION

A substation is a junction point in a distribution or transmission system. Substations are used to:

Switch circuits to control power flow
Isolate faulty sections of the system
Split the system to maintain fault levels
Provide system flexibility
Substations can be one of two types:

1. Air insulated substation (AIS)
2. Gas insulated substation (GIS)

2.2 SUBSTATION ELECTRICAL DIAGRAMS

Substation electrical diagrams includes single line diagrams (SLD) that represent the three-phase wiring diagram by using a single wire diagram. This is because the three phases are similar. You may also come across diagrams that represent protective relays in more detail, called protection SLDs.

2.2.1 Single Line Diagram

A SLD shows the overall relationship between components of electric circuits using single lines and graphic symbols, for example that shown in Fig. 2.1. An example of a protection SLD is shown in Fig. 2.2.

2.2.2 Schematic/Elementary Diagram

A schematic or elementary diagram shows the electrical connections and functions of specific circuit. These can be seen in Fig. 2.3 (AC schematic elementary diagram) and Fig. 2.4 (DC schematic elementary diagram).

Practical Power System and Protective Relays Commissioning.
DOI: https://doi.org/10.1016/B978-0-12-816858-5.00002-2

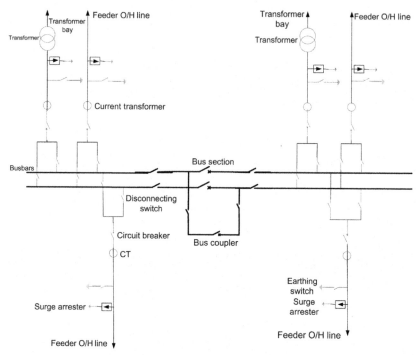

FIGURE 2.1 Substation single line diagram.

2.2.3 Connection or Wiring Diagram

A connection or wiring diagram shows the connections for installation between system components, for example that seen in Fig. 2.5.

2.2.4 Interconnection Diagram

An interconnection diagram shows only the external connections between components of equipment. Refer to Fig. 2.6.

2.3 SUBSTATION AND BUSBAR LAYOUTS

There are several arrangements in which switching equipment can be connected in substations and generating stations. The selection of the arrangement depends on the following:

Degree of flexibility required
Importance of loads
Economic considerations, including availability and cost.

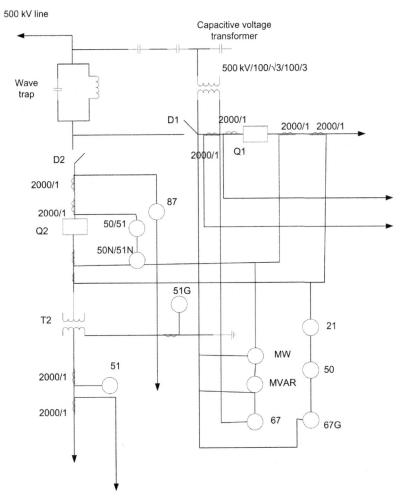

FIGURE 2.2   Protection single line diagram.

Provision of extension.
Protective zones.
Maintenance and safety of personnel.

A discussion of the arrangement of busbars is outlined in the following sections.

2.3.1   Single Busbar Arrangement

This simple arrangement consists of three-phase busbar with various feeders connected to it. Refer to Fig. 2.7.

Total shutdown is required for any maintenance to be undertaken.

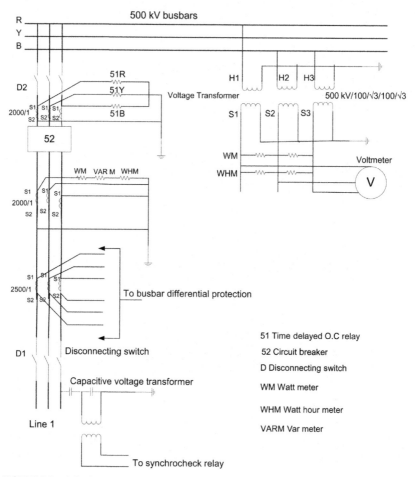

FIGURE 2.3   AC schematic elementary diagram.

2.3.2   Double Busbar/One Circuit Breaker Arrangement

This arrangement is more expensive than a single busbar arrangement but more reliable in maintenance. The circuit can be transferred from the main busbar to the reserve busbar by connecting the bus coupler to make a path for the current in the circuit during the transfer process (Fig. 2.8). The disconnectors cannot be switched off on loads unless there is a path for the current. This scheme is the most commonly used for most utilities.

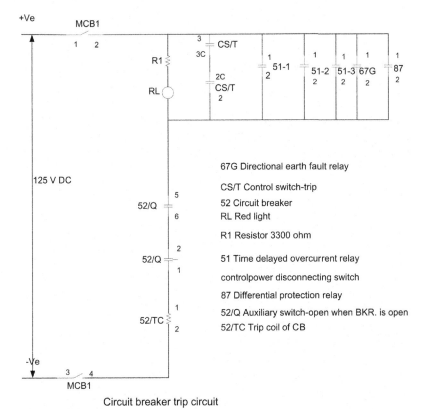

Circuit breaker trip circuit

FIGURE 2.4   DC schematic elementary diagram.

2.3.3   Sectioning of Busbar Arrangement

The busbar can be divided into two sections for more flexibility for maintenance (see Fig. 2.9).

2.3.4   Ring Busbar Arrangement

A ring busbar arrangement provides more flexibility. The supply can be taken from any adjacent section; any fault in one section is localized to that section only (see Fig. 2.10).

2.3.5   One and Half Breaker Arrangement

In this arrangement, three circuit breakers are required for two circuits. This means each circuit has one and half circuit breakers. This methodology is

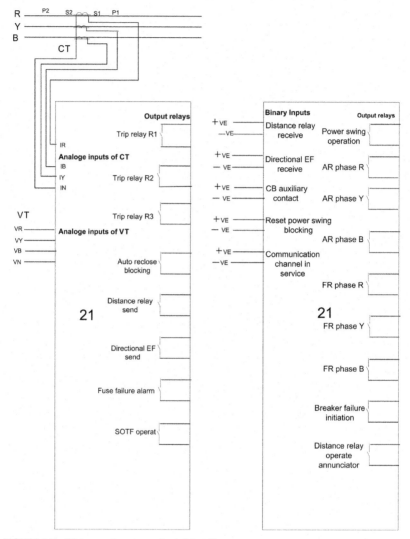

FIGURE 2.5   Distance relay connection/wiring diagram.

more expensive and is used for important high voltage circuits (see Fig. 2.11).

2.3.6   Double Busbar, Double Breaker Arrangement

In this scheme, each circuit has two breakers for each circuit. This is again more expensive but more reliable for very important circuits (Fig. 2.12).

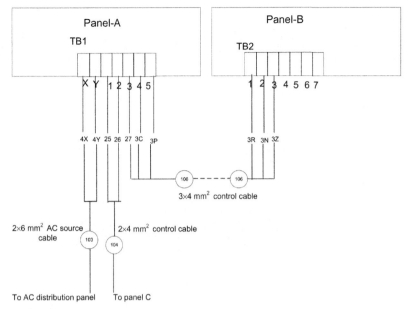

FIGURE 2.6 Interconnection diagram.

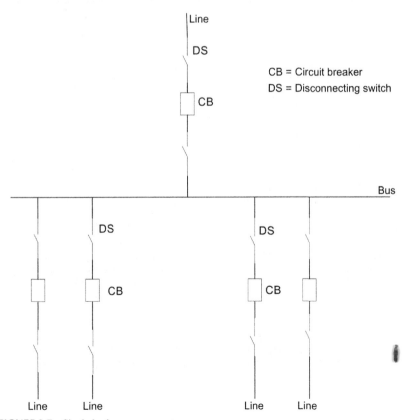

FIGURE 2.7 Single busbar arrangement.

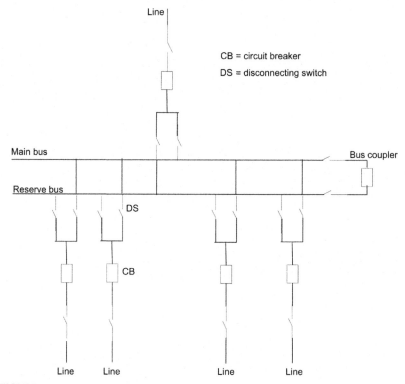

FIGURE 2.8   Double busbar/one circuit breaker arrangement.

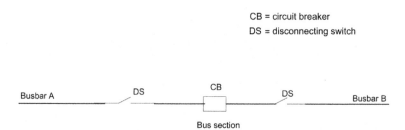

FIGURE 2.9   Busbar in section arrangement.

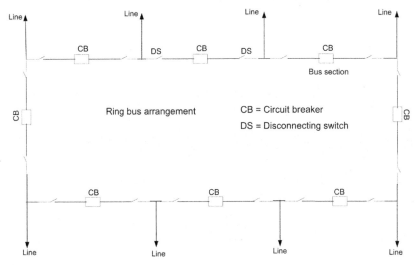

FIGURE 2.10   Ring busbar arrangement.

2.3.7   Interconnected Mesh Corners Arrangement

An interconnected mesh corners scheme gives maximum security against busbar faults (see Fig. 2.13).

2.4   LOAD BREAK SWITCHES

In distribution systems, up to 33 kV are used to switch on loads under normal operating conditions. These switches can be used to make on short circuit current but do not brake on short circuit currents.

2.5   SWITCHGEAR IN GENERATING STATIONS

2.5.1 Main Switchgear Schemes

Classic method of generator connections refer to Fig. 2.14.

2.5.2 Auxiliary Switchgear

2.5.2.1 Unit system of generator

Unit system of generator connections (scheme without C.B for generator) refer to Fig. 2.15.

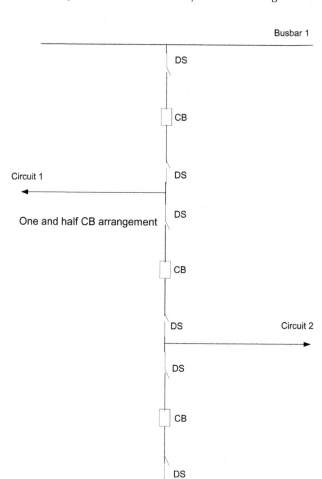

FIGURE 2.11   One and half circuit breaker arrangement.

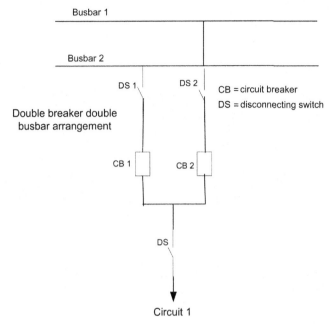

FIGURE 2.12 Double breaker, double busbar arrangement.

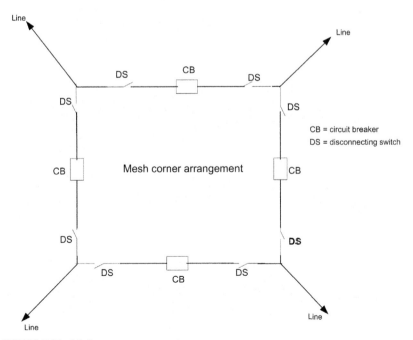

FIGURE 2.13 Mesh corner arrangement.

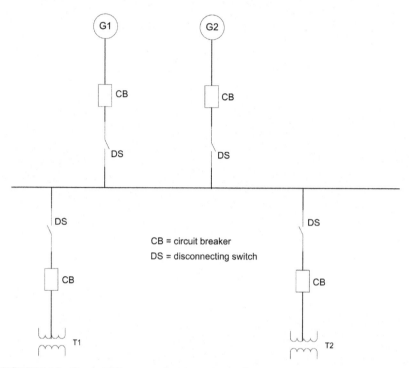

FIGURE 2.14   Classic (old) system of generator connection.

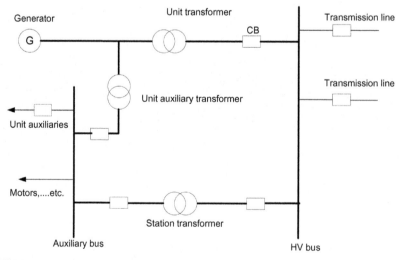

FIGURE 2.15   Unit system of generator connection.

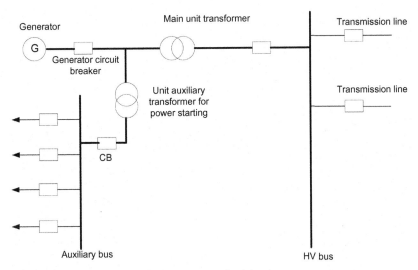

FIGURE 2.16   Unit scheme employing generator circuit breaker.

2.5.2.2 Unit scheme employing generator circuit breaker

Unit scheme employing generator circuit breaker as shown in Fig. 2.16, note that the station transformer is not used, and starting of generator unit auxiliaries is done through the high voltage busbar.

Chapter 3

Introduction to Testing and Commissioning of Power System

3.1 INTRODUCTION

The term "commissioning" refers to a start-up test of the electrical equipment that is undertaken before the initial energization of the equipment.

3.1.1 Precommissioning Procedures

Before commencing a commissioning test, some steps must first be undertaken:

- Approved design drawing should be on site ready to start the work.
- Review all wires in the marshaling kiosk, control panels, protection panels, and low voltage switch boards. After review these wires should be highlighted by yellow marker in the schematic drawings.
- During the wiring review, the tightness of all terminal blocks needs to be confirmed. Low voltage panel bolts tightness torque needs to be reviewed, usually tested at 70 Nm.
- A panel wires review should be undertaken at local control cubicles (LCC) and in protection panels in the control room. The review should also include transformer boxes, low voltage motor control center, and similar items.

3.1.2 Commissioning Management

A commissioning team consists of the commissioning manager and commissioning engineers. The responsibilities of the commissioning manager responsibility are:

1. Producing the commissioning program.
2. Follow up of the commissioning team.
3. Coordinate with other stakeholders on the project regarding commissioning.
4. Ensure that the project has the required resources to perform the different tests required and all material, facilities, tools are on site.

Practical Power System and Protective Relays Commissioning.
DOI: https://doi.org/10.1016/B978-0-12-816858-5.00003-4

5. Producing the switching program and coordinate with national control center (NCC) local/state/federal electrical supply company (also called the National Control Center, NCC) during energization.
6. Producing the commissioning file which includes all method statements, single line diagram, site instructions, deviation list and base design comments, various tests, switching program, relay setting calculations, equipments data sheet and specifications, relay manual, relay testing software, and any other required items.
7. Attend commissioning panel meetings on site with other project members.
8. Produce a contingency plan for an emergency should the required circuits be switched off and the situation requires restoration of the operation of one or more circuit during commissioning of existing circuits modification.

   The responsibilities of the commissioning engineers and technicians are:

   1. Confirm all equipment is installed in the correct manner.
   2. Confirm all equipment and test tools are on site to start activities as per commissioning program (all with valid calibration certificate for at least one year).
   3. Insulation resistance and wiring check for all equipment and panels at site.
   4. Safe working environment at site.
   5. Confirm that the approved relay setting and approved schematic drawings are at site (last revision).
   6. Follow commissioning program steps.
   7. Ensure there is a reliable power source without fluctuation at site to perform the testing in reliable manner; the use of nonstable equipment leads to errors in test results.
   8. Factory test results, approved test procedures and test sheets should be checked by the commissioning engineer on site.
   9. The base and detailed design comments should be checked by the commissioning engineer on site.
   10. Produce a punch list (also known as a snags list) that includes the items not completed in any equipment test in the electrical system.
   11. Produce a marked up drawing on site.
   12. Produce the final commissioning test report.

3.2 POWER TRANSFORMER COMMISSIONING

3.2.1 Visual Check

A visual check of all equipment must be undertaken. These checks should including checking of tightness of all bolts by torque wrench tool, nameplate information, all pipe connections, earthing points, tap changers, and fans.

3.2.2 Electrical Tests

An electrical test can include any of the following:

1. Core insulation resistance.
2. Winding insulation resistance.
3. Winding resistance at all taps.
4. Tan δ of the transformer bushing test.
5. Polarity and vector group tests.
6. Turns ratio test.
7. Tap changer local and remote operation test.
8. Fan groups 1 and fan group 2 (Each transformer has a two groups of fans for cooling) running test for rotation and operation test.
9. Bushing current transformer tests.
10. Oil sample breakdown high voltage test.

These tests are explained in detail in the following chapters.

3.3 SURGE ARRESTOR TEST

Visual checks should include the nameplate, all connections, earthing and all related items as per instruction manual.

3.3.1 Visual Checks

Visual checks for nameplate, connections, earthing and all related items as per instruction manual.

3.3.2 Electrical Tests

An electrical test can include any of the following:

1. Insulation resistance test.
2. Tan δ test.
3. Leakage current measurement.

3.4 CURRENT TRANSFORMER COMMISSIONING TESTS

3.4.1 Visual Checks

Check visually for any damage or craking or oil leakage in the CT.

3.4.2 Commissioning Electrical Tests

An electrical test can include any of the following:

1. Insulation resistance test.
2. Winding resistance test.

3. Polarity test.
4. Ratio test by primary injection.
5. Magnetizing current test.
6. Loop resistance burden test.
7. Continuity of secondary circuits.
   Terminations, shorting links by primary and secondary injection test.
8. High voltage test, which is done during the air insulated substation or gas insulated substation (GIS) high voltage test at the final stages of commissioning tests, but with disconnecting the current transformer (CT) connections to the relays and shorting the CT secondaries during the high voltage test.
9. Demagnetizing the CT cores by connecting the secondaries of the CTs to the ground at the end of all tests. Refer to Chapter 10, Current Transformers, for further details.

3.5 VOLTAGE TRANSFORMER COMMISSIONING TESTS

3.5.1 Visual Checks

Check visually for any damage or craking or oil leakage in the VT.

3.5.2 Electrical Commissioning Tests

An electrical test can include any of the following:

1. Winding resistance test.
2. Insulation resistance test.
3. Ratio test.
4. Polarity test.
5. Miniature circuit breakers (MCBs) trips check by secondary injection in voltage transfer secondary circuits.
6. Phasing test.
7. Loop resistance burden test.
8. Tan δ test.

3.6 GAS INSULATED SWITCHGEAR COMMISSIONING TEST

3.6.1 Visual Check and Mechanical Tests

Checks should include the nameplate information for all components [e.g., circuit breaker (CB), earthing switch (ES), disconnect switch (DS), voltage transformer (VT), current transformer (CT)], enclosure earthing, gas handling unit, mechanical interlocks, and a dew point test for each gas compartment at rated gas pressure of each compartment.

A gas leakage test for all joints by should be undertaken using gas detectors or a plastic bag test around joints. Testing should include the operation of gas density switch alarms and trip signals and lockout functions, and the operation of CBs, DSs, ESs, and LCCs.

3.6.2 Electrical Tests

An electrical test can include any of the following:

1. Insulation resistance test for all wiring and cables.
2. Interlocking circuit test.
3. Busbar joints resistance tests at 100 Amp D.C.
4. Earth resistance at 100 Amp D.C.
5. Function checks of the LCC including bay control unit, interlocking, alarms and interface circuits between the GIS and protection and control circuits.
6. Circuit breaker tests, timing (close and open) DS (close and open) including the special auxiliary contacts of DS used in busbar protection scheme and a CB mechanism check.
7. VT and CT tests.
8. Heater checks in LCC especially in countries with cold weather.
9. High voltage test of GIS components include CBs and DSs, with VT isolated and CT secondary circuited shorted as per international standard for 1 minute.
10. Partial discharge test measurements; refer to Chapter 14, Gas Insulated System Substation.

3.7 HIGH VOLTAGE CABLES COMMISSIONING TESTS

3.7.1 Visual Checks

Checks should include assessing the tightness of connections using a torque wrench tool.

3.7.2 Electrical Tests

An electrical test can include any of the following:

1. Phasing test.
2. HV test.
3. Insulation resistance test.

Refer to Chapter 6, Transmission Lines, for further details.

3.8 PROTECTION AND CONTROL PANELS COMMISSIONING TESTS

3.8.1 Visual Check

Check that the panels are built as per the as-built approved drawing. Other checks should include the earthing of each panel, and all external cables to station equipment.

3.8.2 Electrical Tests

An electrical test can include any of the following:

1. Auxiliary relays tests.
2. Secondary injection tests.
3. Primary injection tests.
4. Scheme function tests.
5. On-load tests.
6. Tripping tests.
7. Metering panel tests.
8. Alarm and annunciators tests.
9. Fault recorder panel tests.
10. End-to-end test for line differential protection and distance schemes.
11. Stability of 380 V tests of transformer differential protection.
12. Remote taps-charger panel tests.
13. Bus transfer (ATS) system in 11 kV switchgear.
14. Scada signal operational and alarms test from station to master station.

3.9 FINAL ENERGIZATION COMMISSIONING PROCEDURES

Before energization these points should be confirmed:

1. Overcurrent commissioning setting is used on bus coupler temporary which will accelerate the relay operation during first time energization and the energization of any circuit will be through the bus coupler of the station.
2. Remove all CT secondary shorting links.
3. All protection relays are working on their final setting.
4. All VT MCBs are closed and no abnormal alarms exist.
5. No engineers or technicians should be in the switchgear area during energization.
6. No earthing exists in the primary circuits.
7. After energizing the new circuit, a phasing test should be done between the circuit VT secondaries and existing reference circuit VT secondaries.

All of this should be carried out before the loading of the new circuit.

After loading of the circuit gradually by the NCC the following should be done:

1. Check the values of each current on each phase for protection relays on each protection panel, including the metering panels.
2. Perform directional tests for directional and distance relays with at least 10%−30% of load current depending on whether it is an electromechanical or digital relay.

3.10   AC DISTRIBUTION PANEL COMMISSIONING TESTS

3.10.1   Visual Checks

A visual check should include the tightness of bolts of the low voltage AC busbars with torque wrench tool. Also checking the MCB and CB ratings as per drawings and checking of panel earthing (correct cable cross-section).

3.10.2   Electrical Tests

An electrical test can include any of the following:

1. Insulation resistance tests.
2. Phasing check for cables.
3. Calibration of meters and ampere and voltage meters.
4. Checking of auxiliary relays.

3.11   DC DISTRIBUTION PANELS COMMISSIONING TESTS

3.11.1   Visual Check

This check should include the MCB ratings and CB ratings as per drawings, the tightness of bolts of busbars with torque wrench tool and checking of panel earthing (correct cable cross-section) for earthing.

3.11.2   Electrical Tests

An electrical test can include any of the following:

1. Insulation resistance of all busbars and internal wiring in the panel.
2. Polarity check.
3. Operation and trip of MCBs.
4. Checking of earth fault relay on this panel.

3.12 BATTERY COMMISSIONING TESTS

3.12.1 Visual Check

A visual check should assess the tightness of bolts by torque wrench tool, check for any earth on the battery's positive and negative lines, the safety equipment in battery room, correct electrolyte specific gravity and liquid level, and the position and working status of the isolation switch of the battery bank.

3.12.2 Electrical Tests

An electrical test can include any of the following:

1. Checking of each cell's voltage.
2. Discharge test.
3. Recharge test.

3.13 BATTERY CHARGER COMMISSIONING TESTS

3.13.1 Visual Checks

Ensure that there are no any damage or disconnected wires inside the panel.

3.13.2 Electrical Tests

An electrical test can include any of the following:

1. Insulation resistance test for external wires.
2. Checking of fast and floating charge operation during battery commissioning tests.

Chapter 4

Generators and Motors: Theory and Testing

4.1 INTRODUCTION

In this chapter we will explain the theory of generators, types, circuit analysis, standard ratings, main components, synchronizations, connections and installation, and the appropriate testing commissioning procedures.

4.2 GENERATING STATIONS

There are different types of generating stations. These generating stations are classified according to the fuel that is used.

4.2.1 Hydro Power Stations

In these generating stations, electricity is produced by water flowing through a turbine that drives an electrical generator. The amount of power produced is proportional to the water head or height of reservoir. The turbine of this machine has a slow speed of 100−250 revolutions per minute (rpm). This machine can produce electricity within a few minutes of startup, which is advantageous for supplying peak loads.

4.2.2 Thermal Power Stations

Thermal power stations generate power from the burning of fuel such as coal, oil, or gas, in boilers of steam generators. Steam generators run at higher speeds of 3000−3600 rpm at 50−60 Hz.

Gas turbine stations also fall into this category. Additionally small size units powered by diesel engines do exist.

4.2.3 Nuclear Power Stations

Nuclear power stations are similar to steam power stations but the heat necessary to change the water into steam comes from the nuclear reactor.

Practical Power System and Protective Relays Commissioning.
DOI: https://doi.org/10.1016/B978-0-12-816858-5.00004-6

4.3 RENEWABLE POWER SYSTEMS

There are two types of energy: primary energy and secondary energy. Primary energy comes directly from the environment, with examples such as nonrenewable energy (fossil fuels) as coal, crude oil, natural gas, nuclear fuel. A secondary energy source is one obtained from a primary energy source through a transformation process, typically with the aim to make it suitable for a particular energy use.

Renewable energy can be categorized into hydro power, biomass, solar energy, wind, geothermal and ocean energy and waste. Primary energy can be converted to secondary energy as petroleum products, electricity, heat, and biofuels.

4.3.1 Introduction

This type of energy is sustainable and coming from nature sources.

4.3.2 Renewable Energy

These sources of energy can be replaced to produce energy and reused by nature without any impact on the environment. Renewable energy sources can produce electricity and now produce up to 20% of the required electricity without any environmental impact. The use of renewable energy has increased in the past 10 years and continues to grow.

4.3.3 Types of Renewable Energy

There are many types of renewable energy sources, some examples are:

1. Solar
2. Wind
3. Geothermal
4. Biomass
5. Ocean or tidal
6. Hydro-electric

4.3.3.1 Solar Energy

Solar electric systems use photovoltaic (PV) systems which utilize sunlight to generate electricity. A PV panel consists of many solar cells. The cells, made from silicon, are grouped into a PV module covered by glass (see Fig. 4.1).

Sunlight is absorbed by the silicon in the PV panel and the PV panel produces a DC current. The DC current is converted to AC current by an inverter as shown in Fig. 4.1. The AC power is injected into the power grid. The solar DC energy can be stored in batteries and these batteries supply loads when there is no light at night and/or when they are fully charged.

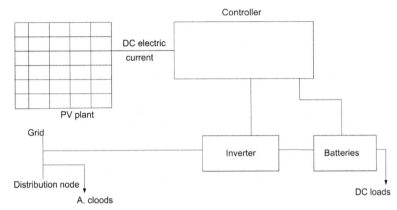

FIGURE 4.1   Photovoltaic system.

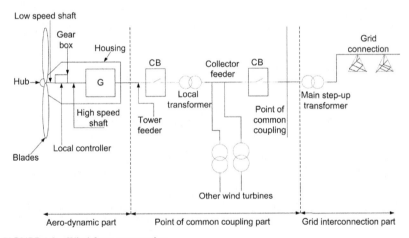

FIGURE 4.2   Wind farm construction.

The system works to supply local loads and inject surplus power into the grid.

4.3.3.2   Wind Energy

This form of energy source uses wind power to generate electricity (Fig. 4.2).

The generator used for a wind farm is a double-feed induction generator. This type of generator cannot provide the power grid with fixed voltage and frequency without the use of a controller.

4.3.3.2.1 Wind Turbine Power System Protection Zones

As shown in Fig. 4.2 there are two protective zones for power system wind turbine. The first zone of protection is the generator zone, which includes the following protections:

1. Generator protection: includes under/over voltage, under/over frequency protection and generator resistance temperature detectors.
2. Generator step-up transformer protection: includes the primary fuses.
3. Circuit breaker (CB) protection: includes overcurrent protection.

The second zone of protection is the collector feeder zone, and includes the following protections:

1. Collector feeder protection: includes overcurrent protection.
2. Collector bus and main step-up transformer protections: this protection has a numerical relay with transformer differential relay, backup overcurrent relay and collector bus differential protection with breaker failure relay.

Lastly, we also need protection of the grid zone. The level of protection of this zone depends on the whether the connection is high voltage or medium voltage.

4.3.3.3 Geothermal Energy

Geothermal energy utilizes the heat buried within the earth for heating houses and producing electricity.

This type of energy is predominantly used for areas known to have earthquakes, volcanoes and hot springs. The steam produced from the geotherms is used to rotate turbines to produce electricity.

4.3.3.4 Biomass Energy

This source of energy is derived from biological sources; any organic material such as wood, wood waste, plants, and animal manure. The heat generated from this material when burned will be used to heat steam to drive a steam turbine to generate electricity.

4.3.3.5 Ocean or Tidal Energy

Ocean or tidal energy is a type of hydro power that converts tides to electricity. Tides are waves caused due to gravitational pull of the moon and the sun.

During high tide, water flows into the dam, and during low tide, water flows out of the dam. These movements cause the turbine to rotate.

4.3.3.6   Hydro-electric Energy

The fall of water onto a turbine (e.g., a natural waterfall or manmade dam) causes a turbine to rotate and produce electricity.

4.4   SYNCHRONOUS GENERATORS: THEORY AND CONSTRUCTION

A synchronous generator consists of a stator with stator winding, and rotor with rotor field winding. When the current passes through the rotor winding, this generates a rotating magnetic field.

This rotating field energizes the stator winding and generates an electromotive force (emf). The emf generates an AC current in the stator winding as shown in Fig. 4.3.

Generator units can have rates between 250 and 800 MW, however higher units do exist with a rated voltage of between 16.5 and 27 kV and rated current of 10,000−20,000 A.

The components of turbogenerators are:

1. Stator: stator frame, stator core, stator winding, end cover, bushings, and generator terminal box.
2. Rotor: rotor shaft, rotor winding, rotor ring, and field connection.
3. Excitation system: pilot exciter, main exciter, and diode wheel.
4. Auxiliaries: bearings, cooling system, and oil supply system.

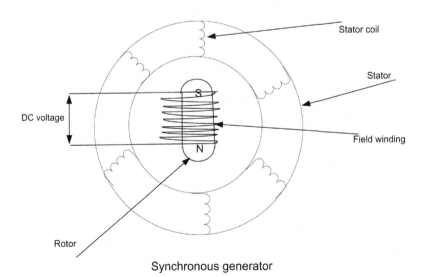

Synchronous generator

FIGURE 4.3   Synchronous generator construction.

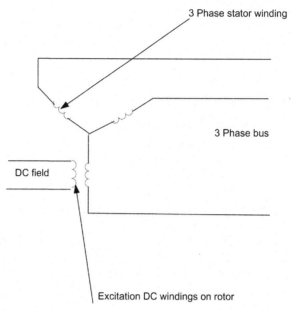

FIGURE 4.4   Synchronous machines.

A synchronous machine is an AC machine whose speed is proportional to the frequency of armature current. The speed under steady state condition is called synchronous speed (Fig. 4.4).

This rotor and stator winding generates three-phase AC power, with an AC magnetic field on a rotor with a DC excitation current feeding the rotor winding. The rotation of rotor this gives a rotating magnetic field. To calculate the synchronous speed (Ns) of the generator, the following equation is used:

$$Ns = \frac{120}{\text{Number of poles(P)}} \times \text{Frequency}(50 \text{ or } 60)\text{Hz}$$

Synchronous generators have damping bars in the rotor to dampen rotor oscillations during the transient condition. The steady-state representation of synchronous generator can be represented as seen in Fig. 4.5.

A series of equations and diagrams can represent the periods that synchronous machines can exist in, as seen in Fig. 4.6:

- Subtransient period with synchronous reactance and field impedance and damping bar impedance in parallel.
- Transient period with synchronous reactance and field impedance only as the current in the damping bar is decaying.
- Steady-state period with synchronous reactance only as the field current is decaying.

$X_{SG \text{ subtransient}}$ $= X_L + (X_a // X_F // X_D)$	$X_{SG \text{ transient}}$ $= X_{L+}(X_a // X_F)$	$X_{SG \text{ steady state}}$ $= X_L + X_a$
X_L = Leakage reactance	R_L = Leakage resistance	R_a = Armature resistance
X_F = Field reactance	R_F = Field resistance	
X_D = Damping bars reactance	R_D = Damping bars resistance	

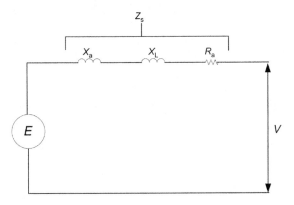

Z_s = Synchronous impedance of synchronous generator(SG)

X_a = Armature reaction reactance of SG

X_L = Leakage reactance of SG

R_a = Stator winding resistance and can be neglected compared to $X_a + X_L$

FIGURE 4.5   Steady-state representation of a synchronous generator.

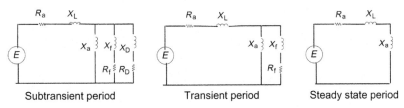

FIGURE 4.6   Synchronous machine reactance in subtransient, transient, and steady-state.

Generator networks generate a large amount of heat. The limitation of generator operation due to heating follows the generator capability curves as shown in Fig. 4.7.

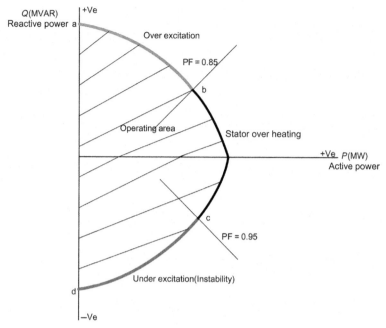

FIGURE 4.7   Generator capability curve.

In Fig. 4.7:

- Region a−b is the overexcitation area
- Region b−c is the stator overheating area
- Region c−d is the underexcitation area
- $+Q$ is the reactive power supplied to the system from the generator
- $-Q$ is the reactive power absorbed by the generator from the system
- PF is the power factor $= \cos(\phi)$.

This curve ensures that the unit will operate in the allowed temperature limits. The Q (MVAR) is controlled in the generator by controlling the excitation field current (I_f) and the P(MW) is controlled by a prime mover (that is, adding more steam to the steam turbine). The equation for this is:

$$MW = MVA \times \cos(\phi)$$

where (ϕ) is the angle between generator terminals voltage and current.

4.4.1   Synchronous Generator Excitation System

A synchronous generator has a field winding on a rotor. This winding needs a DC excitation current. The change in excitation current results in a change of reactive power and a change in the generator terminal voltage. The most used excitation system for generator is the brushless excitation system consists of the following components:

- Pilot exciter
- Main exciter
- Rectifier wheel
- Automatic voltage regulator

In a brushless excitation system a rectifier is mounted on the rotor. The emf induced in the three-phase rotor winding of the main exciter is rectified and supplied to the DC field winding of the main generator on the rotor as shown in Fig. 4.8.

This system does not need slip rings and brushes. It has the main exciter on the main shaft of the generator-turbine with a rotating rectifier and pilot exciter of permanent magnet generator as shown in detail in Fig. 4.9.

This system provides the main generator with a direct current for field winding and controls this current based on the generator output voltage. The

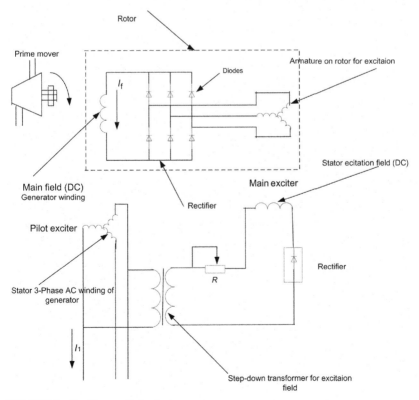

FIGURE 4.8 Brushless excitation system.

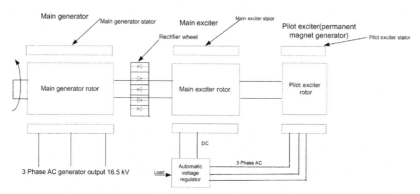

FIGURE 4.9   Brushless excitation system cross-section.

exciter provides DC current to the generator field winding and gives the regulator control of the exciter output based upon the generator output voltage.

4.4.2   Synchronous Generator Ratings

Short-circuit Ratio (SCR) is defined as follows:

$$\text{Short circuit ratio} = \frac{\text{Field current required for rated open circuit voltage}}{\text{Field current required for rated stator Current on SC}}$$

This ratio is inversely proportional to the direct axis synchronous reactance X_d, its typical values lie between 0.5 and 0.6 for turbogenerators and 1.0 and 1.5 for hydro generators. A higher SCR will give a lower synchronous impedance. High short circuit currents and a lower SCR will inversely affect the voltage regulation and stability.

4.4.3   Steady-State Stability and Transient Stability of Synchronous Generators

During a system disturbance, such as a sudden increase in load or sudden switching power swings, the synchronous generator has oscillations of torque angle about the mean position. If a generator has a loss of synchronism due to this disturbance, this called the loss of stability state. If the system can respond to a small, gradual change in power at a given point in the system, this called steady-state stability. The maximum power that can be transferred at a given point in the system with a gradual change in power without loss of synchronism is called the steady state-stability limit. If the system can respond to a big, fast change in power at a given point in the system, this called transient stability. The maximum power that can be transferred at a given point in the system with sudden large change in power without loss of synchronism is called transient-stability limit.

4.4.4   Power Angle of Synchronous Machines

The angle δ between the generator emf voltage vector (E) and generator terminal voltage vector (V) is called a power angle, or a load angle. When the load of a synchronous machine changes, the load angle oscillates about mean position and this condition is called machine hunting. To overcome this condition, a synchronous generator is equipped with a damper winding on the rotor which produces a damping torque.

4.4.5   Excitation Response

The excitation response is the rate of change in exciter voltage and defined by volts per seconds-slope of voltage–time curve at a certain load condition.

4.4.6   Excitation Ceiling Voltage

The maximum voltage that may be given by an exciter at certain load condition is called the excitation ceiling voltage.

4.5   GENERATOR CONNECTION IN POWER STATIONS

As shown in Fig. 4.10 the main items in this scheme of connection are the generator, unit auxiliary transformer, station service transformer, high voltage switchgear and switchgear for transformers with low voltage.

4.6   SYNCHRONIZING OF SYNCHRONOUS GENERATOR WITH BUSBARS

The process of paralleling a synchronous generator with an infinite busbar of a networks with a voltage, phase sequence, and frequency is called generator synchronizing.

Conditions of synchronizing are:

1. The voltage of the generator should be equal to the voltage of the busbar.
2. Phase sequences should be the same for the generator and the busbar.
3. The frequency should be the same for the generator and the busbar

The steps of synchronizing are as follows:

1. Run the synchronous generator in synchronous speed.
2. Increase the excitation current to increase the terminal voltage gradually to the nominal voltage value of generator.
3. Ensure the generator and busbars are at the same voltage and frequency.
4. Check that the phase sequence indicator is anticlockwise.
5. Increase the frequency of the generator to 0.1 cycle more than the busbar frequency.

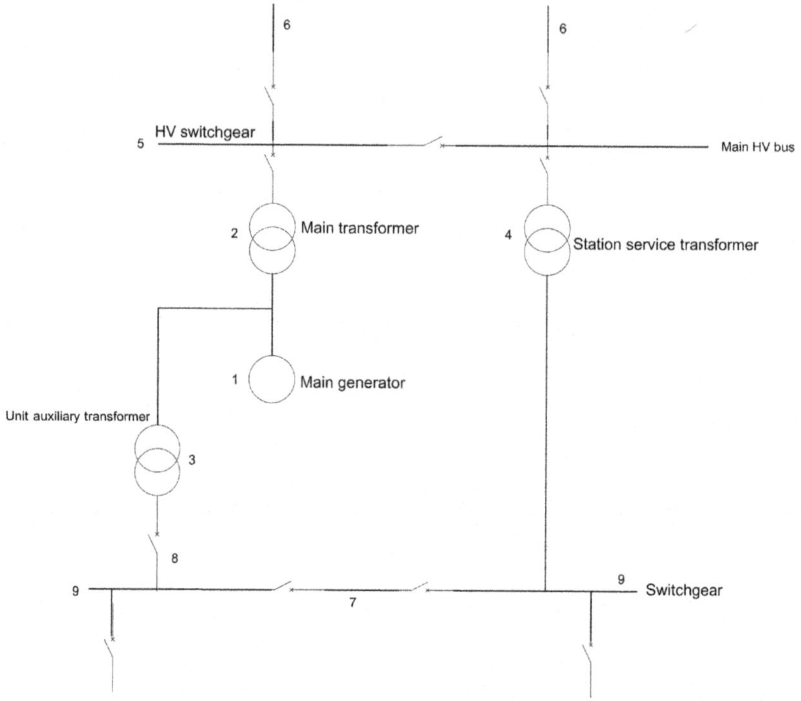

FIGURE 4.10   Generator connection in a power station.

6. Read the synchroscope. If the machine frequency is less than the busbar frequency, the indicator has a "slow" reading, and if the frequency of machine is higher than the busbar frequency, then the machine is giving a "fast" reading.
7. If the synchroscope reading is slow, increase the speed of the machine (synchronous generator) until the pointer is in the middle.
8. When the pointer of synchroscope is in the middle, close the generator CB (zero angle) or at an angle less than 5 degrees in the "fast" direction.

4.7   INSTALLATION AND COMMISSIONING OF SYNCHRONOUS GENERATORS

When the generator arrives on site the following steps should be undertaken:

1. Acceptance and storage at site in a correct manner.
2. The foundation and civil engineering work to be completed as per approved detailed design by the utility engineers.
3. Alignment of turbine and generator shafts to be carried out as per the appropriate international standards.

4. Installation of cooling system of generator to be carried out correctly as per the approved detailed design by the utility engineers.
5. Checking of insulation of generator stator, and supply and control and protection cables by a Megger DC test.
6. Tests on generator stator and other equipment with generator.
7. Drying out of generator: heating up the insulation to reduce the absorbed moisture then repeat the Megger insulation test.
8. Trial run test, no-load run under supervision of the unit and its auxiliaries.
9. Setting and testing of protective relays as per the approved relay setting table and relay testing procedures.
10. Final commissioning and hand over to the customer.

4.7.1   The Pre-start Commissioning Tests of Synchronous Generators

The following prestart commissioning tests should be undertaken:

1. Measurement of the winding resistance of the main stator, exciter stator, and exciter rotor using a micro-Ohm meter.
2. Measurement of insulation resistance of winding. This test is done before and after the high voltage test on the main stator, main rotor, exciter stator, and exciter rotor. A Megger 1000 V is used for stator winding for 1 minute. The Polarization Index Test is used to calculate the ratio of insulation resistance for 10 minutes through to insulation resistance for 1 minute.
3. High voltage tests at power frequency on main stator. This test is done on site for generator stator winding under a voltage following this equation:

   V test $= (2\,V + 1)$ kV between phase and ground with the other two phases connected to the ground for 1 m for each phase. The test voltage is raised quickly to the maximum test voltage. It is kept at that value for 1 minute and then reduced slowly to zero.
4. High voltage tests of the power frequency on the main rotor, exciter stator, and rotor. This test is done on site under a voltage following this equation:

   V test $= (10 \times$ field voltage/nominal voltage) Volts not less than 1500 voltage AC between phase and ground with the other two phases are connected to the ground for 1 minute for each phase.
5. Open circuit characteristic test. The generator is run at a rated speed in an open circuit gradually increase the excitation voltage to have 20%, 40%, 60%, 80%, 90%, 95%, 100%, 105%, 110%, and 120% of rated voltage. Record the generator voltage, excitation voltage, and excitation current then create a graph with the open circuit

characteristic field current shown on the X axis and the rated voltage on the Y axis.

6. Short circuit characteristic test. Run the generator at the rated speed in the short circuit of generator stator terminals then record the generator short circuited current at 25%, 50%, 75%, 100% of excitation voltage and current. Create a graph with the short circuit characteristic field current on the X axis and rated current on the Y axis.

7. Phase sequence test with respect to direction of rotation. Measure the phase sequence in the rotation direction of anticlockwise.

8. Voltage balance test. The three-phase voltages are checked for balance when generator runs in the rated voltage and rated speed but in an open circuit condition.

9. Vibration test. Run the generator in the rated voltage and rated speed; the vibration is measured at the vertical, horizontal, and axial direction on the bearings.

10. Impedance test of rotor. One-phase variable voltage is applied to rotor field winding and current is measured. Record the reading at 25%, 50%, 75%, and 100% of the rated field voltage.

11. Momentary overload test. The generator is run at the rated speed with stator terminals short-circuited, then generator is excited to 150% of its field current for 30 seconds.

12. Noise level test. With the machine running at the rated speed, a voltage measurement on no-load and noise is taken at different locations of the machine, distance 1 m, using a noise level meter. Then the machine is switched off and ambient noise is measured. Using these two numbers, the machine noise is calculated.

13. Over speed test. The machine is run at 120% of the rated speed for 2 minutes. No disturbance should be noticed during that time.

14. Wave form measurement test. The machine is run at the rated speed and rated voltage; the three-phase voltages are recorded using a suitable instrument.

15. Trial tests.

16. On load tests.

17. Tests of the cooling system.

18. Testing of the lubrication system.

4.8 SYNCHRONOUS MOTORS THEORY

Synchronous motors consist of stator and rotor windings that are placed on the stator and the rotor respectively as shown in Fig. 4.11.

The stator winding supplied with a three-phase supply generates a rotating field in the stator. The two rotating fields in the stator and the fixed field on the rotor are linked and cause the rotor to rotate in a synchronous speed by a torque generated by the two electric fields.

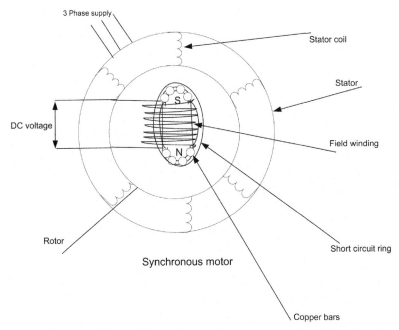

3 Phase supply

Stator coil

Stator

DC voltage

Field winding

Rotor

Short circuit ring

Synchronous motor

Copper bars

FIGURE 4.11   Synchronous motor construction.

To start the motor, we use a small DC motor to rotate the rotor with synchronous speed, then the field DC supply is connected. An alternative method is to use copper bars with a shorting ring to generate a field that rotates the rotors with synchronous speed, then the DC supply is connected to the field winding on the rotor. These bars dampen the change in load angle of the motor to keep the motor synchronized.

4.9   THREE-PHASE INDUCTION MOTORS THEORY

The stator in connected to a three-phase supply to generate a rotating field that rotates with synchronous speed. This electric field cuts the rotor field winding which is short-circuited and hence generates a rotating electric field on the rotor rotates (Fig. 4.12). Its speed (n) with a slip from synchronous speed S is calculated by: $S = (ns - n)$ where ns = synchronous speed, n = motor speed, and s = slip to synchronous speed of the stator rotating field.

The starting current of the induction motor is higher than the full load current as per the following equation:

$$\frac{I\text{start}}{I\text{full load}} = (3{:}8)$$

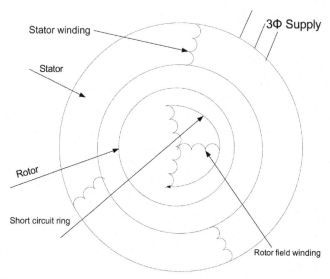

FIGURE 4.12 Induction motor construction.

For starting the induction motor there are a different methods some of the common used one are as follows methods:

1. Rotor resistance method.
2. Stator resistance method.
3. Auto-transformer method.
4. Star–delta method.

4.10 THE PRE-START COMMISSIONING TESTS OF INDUCTION MOTORS

The following prestart commissioning tests should be undertaken:

1. Insulation resistance measurement. This test is done by using a Megger test of phase-to-earth and phase-to-phase to check the dryness and quality of insulation.
2. Measurement of winding DC resistance, to check the winding and circuits.
3. High voltage tests and insulation tests. This test is done to check the quality of insulation.
4. Phase sequence check
5. Tests of protective relays. This test is done to check the protective relays at the approved setting from the client's engineers.
6. Tests of motor auxiliaries and motor cooling system.

7. On load tests.
8. Trial run test under supervision.
9. Motor performance characteristic test.
10. Vibration tests. These tests are done under no-load and full-load conditions, and a trial run condition.

Chapter 5

Power Transformers Theory Testing and Commissioning

5.1 INTRODUCTION

A power transformer is the device that transfers the power through power system with different voltage levels. Both step-up power transformers and step-down power transformers exist. This is a very expensive component in a power system and should be maintained in good condition to perform at the optimal level. A new device is introduced to monitor the insulation condition by monitoring the partial discharge in transformers as live monitoring and can help to predict when the transformer insulation will fail, which is the most common cause of failure of power transformers.

5.2 TRANSFORMER CONSTRUCTION

Fig. 5.1 shows the different parts of a transformer.

A transformer consists of the tank and its accessories, core, and winding assembly, and the insulating and cooling medium. The tank is made from welded steel and painted, and is mechanically very strong in able to support the transformer core and windings and the weight of the oil. Accessories may include the following:

- A breather to compensate for the expansion and contraction of the cooling medium (oil)due to temperature changes. Small units have an air space above the oil connected to the atmosphere by the breather. The breather should protect the oil from changes due to humidity, snow, etc.
- A conservator or expansion tank it is a small tank that keeps the main tank full of oil and allows for the expansion and contraction of oil due to temperature changes and has a breather connected to the atmosphere.
- A pressure relief valve which relieves the high pressure that occurs inside the transformer under any electrical fault condition.
- Temperature indicators that include:
  - Oil temperature indicator set in two stages: alarm, and starting fans for cooling
  - Winding temperature indicator set in two stages: alarm and trip.

Practical Power System and Protective Relays Commissioning.
DOI: https://doi.org/10.1016/B978-0-12-816858-5.00005-8
49

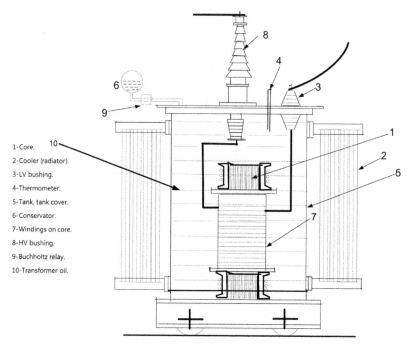

1-Core.
2-Cooler (radiator).
3-LV bushing.
4-Thermometer.
5-Tank, tank cover.
6-Conservator.
7-Windings on core.
8-HV bushing.
9-Buchholtz relay.
10-Transformer oil.

FIGURE 5.1 Parts of a power transformer.

- The Buchholz relay is a relay with two floats or buckets; one for alarm and the other to trip the transformer (see Fig. 5.2).
- Bushings to isolate the electrical circuit of the power transformer from the tank. This is usually composed of an outer porcelain body that is suitable for high voltages, and also an oil-impregnated paper as insulation within the porcelain column of bushing.
- Pumps, fans, and radiators as a part of the cooling system of transformer.
- Vacuum valves.
- Liquid handling valves and sampling valves.

5.3 TRANSFORMER COOLING SYSTEM

When discussing transformer cooling systems, the following terms are used:

- Internal cooling medium oil, referred to as (O)
- External cooling medium air, referred to as (A), or water, referred to as (W)
- Type of circulation, referred to as natural (N) or forced (F).

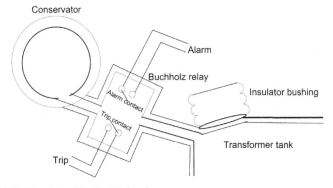

FIGURE 5.2 Principle of the Buchholz relay.

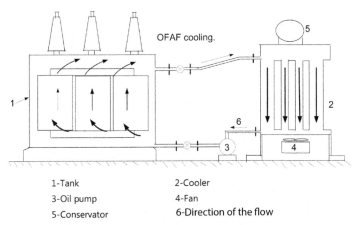

1-Tank   2-Cooler
3-Oil pump   4-Fan
5-Conservator   6-Direction of the flow

FIGURE 5.3 OFAF transformer cooling system. *OFAF*, Oil forced air forced.

Refer to Fig. 5.3 for a diagram of oil forced air forced (OFAF) cooling and Fig. 5.4 for oil forced water forced (OFWF) cooling. Table 5.1 shows transformer cooling codings and an explanation of their methods.

5.4 TRANSFORMER THEORY AND CIRCUIT ANALYSIS

A power transformer transfers energy from one voltage level to another voltage level, consisting of two or more coils of wires wound around a magnetic core (see Fig. 5.5).

Transformer impedance can be represented in percentage impedance or per unit value. The impedance represents the losses in the transformers due to leakage flux (defined as magnetic flux that does not follow the particularly intended path in a magnetic circuit).

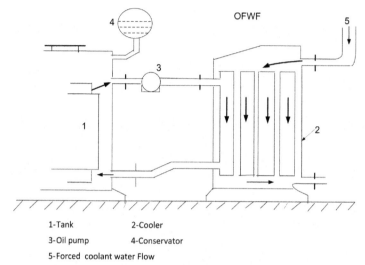

1-Tank   2-Cooler
3-Oil pump   4-Conservator
5-Forced coolant water Flow

FIGURE 5.4 OFWF oil and water are circulated in closed loop by pumps. *OFWF*, Oil forced water forced.

TABLE 5.1 Transformer Cooling Methods

Symbol	Title	Description
ONAN	Oil natural Air natural	Transformers are radiators for oil natural circulation
ONAF	Oil natural Air forced	The radiators are provided with cooling fans. Fans are switched on in two stages
OFAF	Oil forced Air forced	The oil is circulated through coolers by pumps. The coolers have cooling fans to exchange heat from oil into air
OFWF	Oil forced Water forced	The heat is exchanged from oil to cooling water. Both oil and cooling water are circulated through radiator by pumps
AN	Air natural	The ambient air is used for cooling

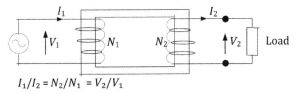

$$I_1/I_2 = N_2/N_1 = V_2/V_1$$

N_1 = Number of turns in primary winding

N_2 = Number of turns in secondary winding

V_1 = Primary voltage

V_2 = Secondary current

FIGURE 5.5   Transformer.

FIGURE 5.6   Percentage impedance of transformer.

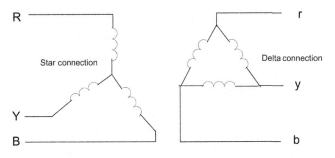

FIGURE 5.7   Star/delta transformer connection.

A transformer is represented in electrical diagrams as a series impedance connected between high and low voltage terminals (Fig. 5.6).

Three-phase transformers can be connected in different ways, for example, in star/delta connection as seen in Fig. 5.7.

Tertiary (third) winding can be used in power transformers for supplying station services, however the disadvantage is that any fault will trip the power transformer.

Auto-transformers are a special type of transformer design. In this design the coils are wound around an iron core as shown in Fig. 5.8.

In a transformer tap changer, a change of voltage is effected by changing the number of turns in the transformer high voltage side so the current level is lower than the low voltage side, with the change undertaken when the transformer is under load. When in tap changing condition, the load should

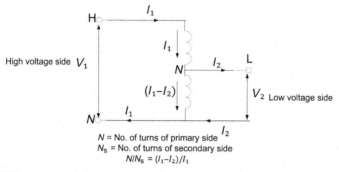

FIGURE 5.8   Auto-transformer.

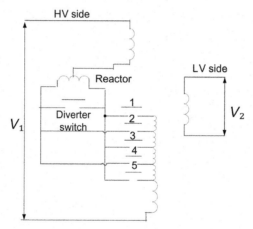

FIGURE 5.9   On-load tap changer using a reactor.

not be broken to prevent damage of the contacts. Additionally, no part of the winding should be short circuited during adjusting the taps as the contacts and the short circuited part of the winding may be damaged. To do this, we use a resistance or inductive reactance to limit the current and a diverter or parallel circuit to carry the load current during the tap changing process (see Figs. 5.9 and 5.10).

Poly-phase transformers are described using symbols that indicate the type of phase-connection and the angle between the primary and second-ary voltages. The angle is indicated by a clockface hour figure with the high voltage vector being set at 12 o'clock (zero) and the corresponding low voltage vector being represented by the hour hand. As an example, in the term "Yd11" , the Y indicates a high voltage star connection, the *d* indicates a low voltage delta connection, and 11 o'clock means +30 degrees in advance of 12 o'clock position of the high voltage. The groups

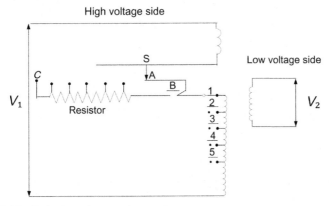

FIGURE 5.10   On-load tap changer using a resistor.

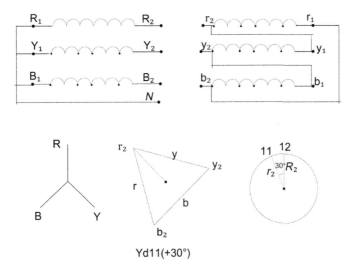

Yd11(+30°)

FIGURE 5.11   Three-phase power transformer connection vector group (Group 4 with 30 degrees lead).

of all possible three-phase transformer connections are categorized as follows:

Group 1: 0 degree phase displacement (Yy0, Dd0, Dz0).
Group 2: 180 degrees phase displacement (Yy6, Dd6, Dz6).
Group 3: 30 degrees lag phase displacement (Dy1, Yd1, Yz1).
Group 4: 30 degrees lead phase displacement (Dy11, Yd11, Yz11).

See also Figs. 5.11−5.13.

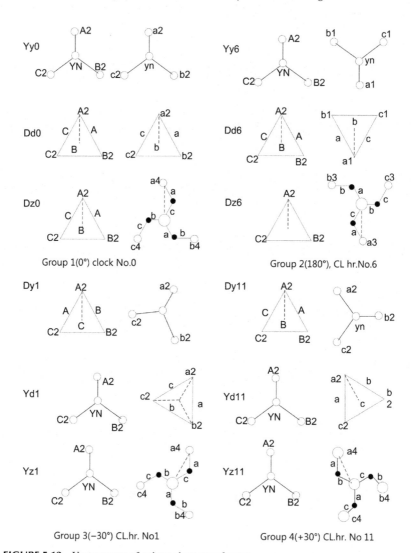

FIGURE 5.12 Vector group of a three-phase transformer.

5.4.1 Transformers and Overexcitation

The design of a transformer is intended to keep the flux density to 1.8 Tesla (T). During operation, however, if the transformer is subjected to more flux density, then this condition is called overexcitation. Overexcitation of flux should not exceed 110%.

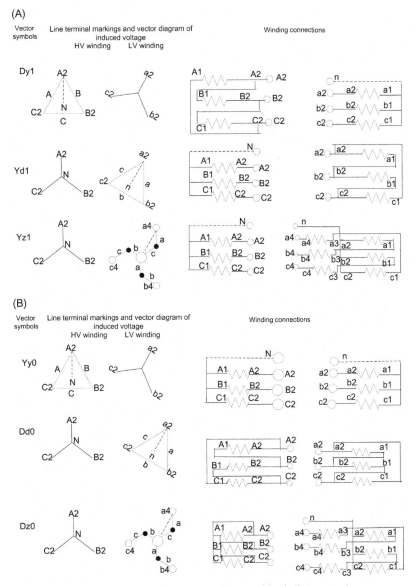

FIGURE 5.13 Vector groups of three-phase transformers with winding connections.

The main causes of overexcitation are:

1. System over-voltage.
2. System underfrequency.
3. Light loading of TL (Transmission Line).

Overfluxing causes core saturation and this causes heating. Over heating of the transformer winding can cause gases which affect the operation of the Buchholz Relay.

During saturation of the transformer core, a rise in the primary magnetizing current can cause operation of differential relay. This can be detected by fifth harmonic blocking of the differential relay.

Overexcitation can be determined by using the V/F ratio. When this ratio exceeds 1.4 then the overexcitation relay trips the transformer instantly.

5.5   POWER TRANSFORMER INSTALLATION

5.5.1   Transportation

Transportation of a power transformer depends on its weight. The transformer is filled with oil above the winding by $15-20$ cm, the oil transported with nitrogen with a pressure of 0.1 bar. The transformer bushing, as well as the current transformers, are also filled with oil.

5.5.2   Site Inspections

Once on site, the following checks of the transformer should be made:

Check the tightness between the transformer and the base which was transported with it.

Inspect the transformer tank, pumps, oil seals, and valves to see if they are correct and there is no oil leakage.

Inspect the condition of transformer bushing and current transformers.

Inspect the cooling system of the transformer.

An oil sample should be taken from the bottom of the transformer and sent to a laboratory for sample chemical analysis.

The transformer silica gel should be blue in color. If there are a lot of brown particles, then the transformer should be dried and tested.

5.5.3   Transformer Storage

A transformer should be filled with oil maximum of one month from the date of it reaching the site, if it did not contain oil during transportation.

The transformer silica gel should be checked and replaced if it is not of optimal condition.

Current transformers should be stored in oil with minimum breakdown voltage of 30 kV.

The transformer bushing should be filled with oil during storage.

5.6   POWER TRANSFORMER TESTING AND COMMISSIONING

5.6.1   Measurement of Winding Insulation Resistance

In this test we use a Megger tester for insulation measurements 5 kV for 1 minute; if the reading is not stable then we should extend the test for 5 minutes.

As a reference guide, when testing a 66/11 kV star/delta power transformer the following results were derived:

3 phases 66 kV winding to earth is 5000 MΩ.
3 phases 11 kV winding to earth is 1000 MΩ.
3 phases 66 kV winding to 11 kV winding is 1500 MΩ.

Refer to Fig. 5.14.

The insulation resistance measured for 1 minute is called R_{60}. It is also measured for 15 seconds at the site temperature. These two calculations should be compared to the factory temperature as shown in Table 5.2.

The calculation should be:

R_{60} measured at factory $= K \times R_{60}$ measured at site

$$\text{Insulation difference} = \frac{R_{60}\text{measured} - R_{60}\text{at factory}}{R_{60}\text{at factory}}$$

$\times 100$ should be less 30%

This percentage for overhaul maintenance should not exceed 15%.

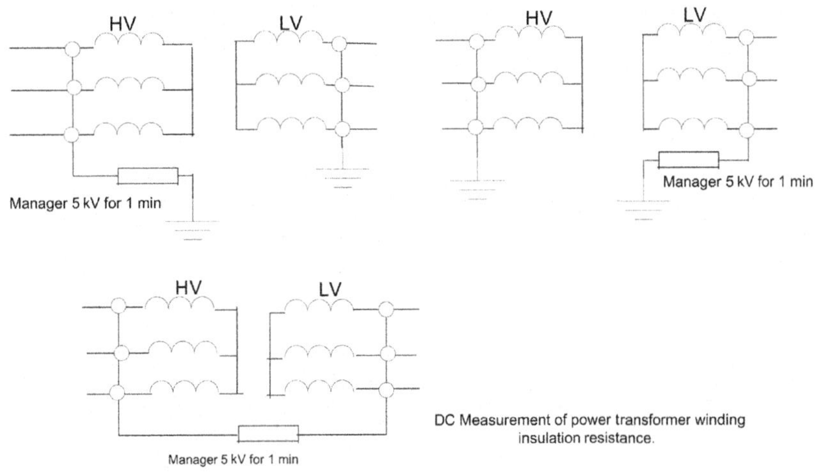

FIGURE 5.14   Measurement of power transformer winding insulation resistance.

TABLE 5.2 Temperature Difference Coefficient K

Temperature Difference	1	2	3	4	5	10	15	20	25	30
Coefficient K	1.04	1.08	1.13	1.17	1.22	1.5	1.84	2.25	2.75	3.4

The ratio R_{60}/R_{15} should not be less than 1.3 for any transformer with a voltage level ≤ 35 kV, and $1.5-1.7$ for transformers with voltage levels of 66, 132, 220, and 500 kV.

As an example:

$R_{60} = 520$ MΩ at 57°C measured at site.
$R_{60} = 510$ MΩ at 55°C measured at factory.
Temperature difference $= 57-55 = 2$°C. Using Table 5.2, the coefficient k is equal to 1.08.
R_{60} measured corresponding to factory temperature $= 1.08 \times 520 = 561.6$ MΩ.
Insulation difference $= (561.6-510)/510 \times 100 = 10.11\%$, which is less than 30%
After the test is finished the transformer winding should be earthed to ensure that it is discharged.

5.6.2 Measurement of Voltage Ratio Test

This test is undertaken by injecting 380 V on site and the ratio is measured at each tap of the tap changer. The transformer should not be tested if it is under vacuum or if there is any doubt about any short circuit in the transformer as this may lead to a flashover and can cause fire or explosion.

The equation is:

Voltage ratio $(K) =$ (high voltage side line to line voltage/low voltage side line to line voltage).

This test can be done using an accurate instrument to measure the voltage or by using a ratiometer.

Regard Fig. 5.15. The difference between the readings of this ratio at site and the factory readings should not exceed $\pm 2\%$, otherwise there is a possibility of a wrong connection in the tap changer or a short circuit in the transformer windings. After the test is completed, the transformer winding should be earthed to ensure that it is discharged.

5.6.3 Determination of Transformer Vector Group Test

The vector group can be determined by making a jumper between the high voltage phase R and the low voltage phase r and injecting a three-phase

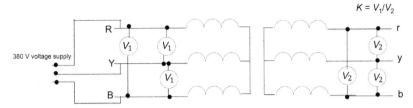

FIGURE 5.15   Voltage ratio test of power transformer.

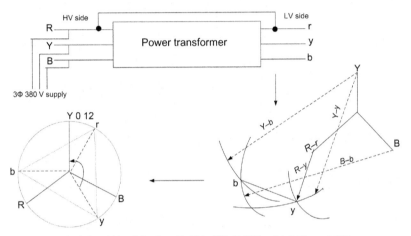

Low voltage (y) lags high voltage side (Y) by 150° = 5 x 30° then the vector group is Yd5

FIGURE 5.16   Vector group test of power transformer.

power supply of 380 V from the high voltage side and measuring the voltages between $Y-y$, $B-y$, $B-b$, and $Y-b$. These voltage vectors are then drawn, noting that the voltages R and r should coincide as they are at the same potential and phase displacement as there is a jumper between R and r. An example vector diagram is shown in Fig. 5.16.

Then next step is to transfer the vector diagram to the clock remembering that the phase Y are on the 0 degree or 12 o'clock position. We find that the voltage (y) is lagging the voltage (Y) by 150 degrees (5 multiplied by 30 degrees) and the vector group is nominated as (Yd5). After the test is complete, the transformer winding should be earthed to ensure that it is discharged.

5.6.4   Winding Resistance Test

In a winding resistance test we use a voltmeter and an ampere meter (also known as an ammeter) and a battery of 12 or 24 V. For low-resistance values we connect the voltmeter after the ammeter, but in high-resistance values we connect the voltmeter before the ammeter. A heavy current of 100 A and sensitive meters should give good results.

During the test we connect the voltmeter after the current reaches the saturation point and we disconnect voltmeter before disconnecting the ammeter. To reduce the time needed for the current to reach saturation condition, we connect a variable resistance in series with the circuit.

The resistance reading at site should be compared to the resistance temperature reading at factory by using this equation:

$$R_2 = \frac{(T_2 + 235) \times R_1}{(T_1 + 235)}$$

where R_1 is the Resistance reading at factory, T_1 is the Temperature which reading is taken at factory, R_2 is the Resistance reading at site, T_2 is the Temperature which reading is taken at site.

The measurement difference, D, calculated by

$$D = (R_2 - R_1)/R_1 \times 100$$

should not exceed 2%.

If the resistance measurement is very low, there is a possibility of short circuits in the transformer windings.

If the resistance measurement is low, then there is a possibility of bad contacts between the winding and insulators, or bad contact in the tap changer, or the winding may be damaged.

5.6.5 Measurement of No-Load Current and No-Load Circuit Losses Test

Under open-circuited conditions on a transformer's high voltage side, the transformer no-load current (magnetizing current) is less than 4% of full load current. Any power losses in this condition is the loss of iron which includes hysteresis losses and eddy current losses. The test is done as shown in Fig. 5.17.

P_0 = Measured no-load power losses

No-load current losses $I_0 = A$—ammeter reading in each phase.

These losses affect the power transformer efficiency.

5.6.6 Measurement of Load Losses: Current and Impedance Voltage Test

Short-circuit (load) losses and short-circuit impedance voltage are guaranteed and reported values by the manufacturer to customers. Short-circuit impedance voltage is an important parameter especially for the parallel operation of the transformers, whereas short-circuit losses are important from economical point of view as they effect the transformer efficiency.

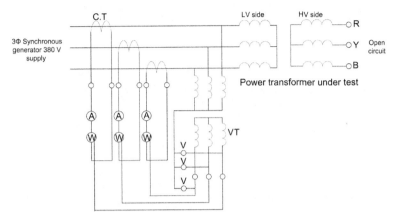

FIGURE 5.17 No-load current and power losses measurement.

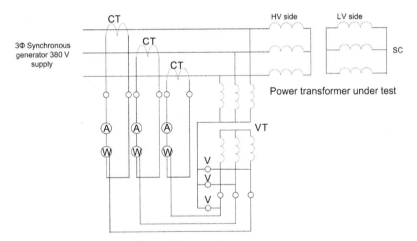

FIGURE 5.18 Load losses, current measurement, and voltage impedance measurement.

The secondary winding low voltage side is short-circuited and the secondary current is circulated in the circuit. Power losses represent the total I^2R losses in primary and secondary windings, and also include stray losses.

According to the Standards, the measured value of the losses is evaluated at a reference temperature (e.g., at 75°C). The measuring temperature (t_m) losses are corrected to the reference temperature (75°C) according to the Standards.

The voltage applied to the transformer to circulate the rated current in the short-circuited secondary winding is called impedance voltage of power transformer.

Refer to Fig. 5.18 for the test circuit.

5.6.7 Temperature Rise Test

The temperature rise test is undertaken using a back-to-back test of two parallel transformers. In most substations there are more than one transformer of the same rating, so this test is possible. In a parallel operation the following conditions should exist:

- Polarity should be the same.
- The voltage ratio should be the same.
- The percentage impedance should be equal.
- Phase rotation should be the same.
- The two transformers should have the same vector group.

In back-to-back testing, the tap of one transformer should move in the rise direction and the other in a lowering direction until the voltage difference between the two transformers are driving the nominal load current. A graph is drawn to show the relation between temperature and time (24 hours) for the oil transformer and the transformer winding temperature. The results should be compared with the results from the manufacturer.

5.6.8 Transformer Oil Breakdown Test

In a new transformer, the oil should be a clear yellow color. A black or dark-color oil means that the oil cannot be used. Impurities have a bad effect on transformer oil and its insulating characteristics, which can lead to flashover. Humidity also reduces oil insulation. A high flash point of 145°C is preferable. During this test, the oil humidity will be measured and the sample is tested in standard cells of 80 mm × 60 mm × 100 m (length × width × height). The poles are two yellow copper balls with a distance between the balls of 4 ± 0.02 mm. The sample of oil is taken from the bottom of the transformer. Good oil will be workable up to 45 kV for 1 minute. Repeat the test at least five times to allow bubbles to escape.

5.6.9 Measurement of Capacitance and Tan δ

All insulating materials used in practice have slightly small dielectric losses at the rated voltage and at the rated frequency. These losses are fairly low for good insulating materials. This loss changes proportionally with the square of the applied voltage. Insulation, in terms of basic circuit elements, is shown in Fig. 5.19.

As it can be seen from Fig. 5.19C the angle δ between the total current "I" and capacitive current "I_c" is a basic value. Insulation angle is dependent on the thickness, the surface, and the properties of the insulation material

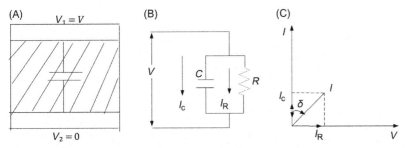

FIGURE 5.19 (A) The object under test, (B) The equivalent circuit of the object under test, (C) The vector Diagram.

TABLE 5.3 Temperature Difference Coefficient C

Temperature Difference	1	2	3	4	5	10	15	20	25	30
Coefficient C	1.03	1.06	1.09	1.12	1.15	1.31	1.51	1.75	2.00	2.3

(the pores, impurities, and humidity that cause the ionization in the insulation material).

The power factor measurement of the insulation at certain frequency gives basic information regarding the insulation. The measurements to be made during the service are one of the most important indications, indicating the age of the insulation and the contamination of the oil. The active losses of the circuit are calculated as follows

$$P = V \cdot I \cdot \cos \phi = V \cdot C \cdot \omega \cdot \tan\delta$$

Capacitance, tan δ, active losses, and cos ϕ can be measured by bridge methods or a power factor (cos ϕ) measuring instrument at definite voltages.

Measurement is performed between the windings and the tank, and the test temperature is recorded; then according to desired reference value the necessary corrections are undertaken. Most test devices use 10 kV as a test voltage but there is a correction factor C to refer the reading to the temperature measured at factory as per Table 5.3.

tan $\delta1$ should be measured at the factory, and tan $\delta2$ measured at site, and the calculation is:

$$\text{Difference } (D) = \frac{\tan \delta1 - \tan \delta2}{\tan \delta1} \times 100$$

Difference (D) should be less than 30% and less than 15% in overhaul maintenance. Refer to Fig. 5.20 for the equivalent circuit of the electrical insulation of two winding transformers.

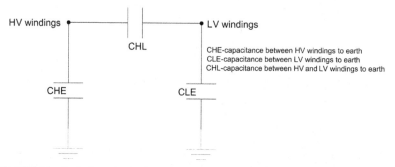

FIGURE 5.20   Equivalent circuit of the electrical insulation of two winding transformer.

As a guide, a typical value of tan δ will depend on the transformer age as follows:

- CHE = 0.13% for new transformers and 1.5% for old transformers
- CLE = 0.15% for new transformers and 1.5% for old transformers
- CHL = 0.2% for new transformers and 1.5% for old transformers

5.6.10   Frequency Response Analysis

Frequency response analysis-FRA- or what we call in the field of transformer testing (SFR test)- sweep frequency response analysis -can determine if the windings of a transformer have moved or shifted. It can be completed as a factory test prior to shipment and repeated after the transformer is received onsite to determine if windings have been damaged or shifted during shipping. This test is also helpful if a protective relay has tripped or a through fault, short circuit, or ground fault has occurred. A sweep frequency is generally placed on each of the high voltage windings, and the signal is detected on the low-voltage windings. This provides a picture of the frequency transfer function of the windings. If the windings have been displaced or shifted, test results will differ markedly from prior tests. Test results are kept in transformer history files so they can be compared to later tests. Results are determined by comparison to baseline or previous measurements or comparison to units of similar design and construction.

5.6.11   Partial Discharge Measurement

The purpose of this test is to measure the partial discharges in the tested object produced by the application of AC voltages during the tests. This test gives comprehensive information about the quality of the insulating materials and the design. The partial discharge is measured to:

Determine the existence of a definite partial discharge in the tested object at a predetermined voltage. This is done by increasing the applied voltage where the partial discharging begins (partial discharge inception voltage) and by decreasing the applied voltage where the partial discharging extinguishes (partial-discharge extinction voltage).

To determine the magnitude of the partial discharge at a predetermined voltage.

The partial discharges (which do not cause flashover between the electrodes) are the discharges in a certain area of the insulation between the conductors of the test object. These discharges may occur in the gaps of the insulating environment, in the gaps of the solid-materials or in the contact surfaces of two different insulations. This discharge can be captured as a single current impulse in the outer region. Although these discharges do not cause permanent deterioration in the insulating media as their energy is relatively small, the thermal energy of the discharges shall cause depreciation, aging, and deterioration in the insulating media. The electrical discharge magnitude at the partial discharge point is not a direct measurement for deterioration of the insulating material in this region. Besides the numerical value, the intensity and the waveform of the impulse, regional discharge concentration, the manufacturing and the placing of the insulation also affects the situation.

5.7 COMMISSIONING TESTS FOR POWER TRANSFORMERS AT SITE

5.7.1 General

Testing of a complete power transformer includes:

Winding insulation resistance, winding resistance, vector group, tan delta, no-load and power losses test, short-circuit test and copper losses, temperature rise test and transformer oil breakdown test.

Control and relay panels testing: junction boxes (local control panels) and marshaling kiosks (including transformer and fan control box and tap changer box).

Transformer fans and pumps.

Check the operation of the tap changer with a voltage regulator relay during secondary tests.

5.7.2 Primary Tests

Stability test of transformer differential relays using 380 V (explained elsewhere in this book).

Primary current injection in transformer bay to check the circuit continuity in secondary circuits.

Stability test of busbar protection at the same time. These latter two items will be explained elsewhere in this book.

5.7.3 Secondary Tests

Secondary injection of all transformer protection relays.
Check voltage regulator relay with the tap changer operation.
Check all alarm circuits.

5.7.4 Tripping Tests

All tripping matrix of high voltage circuit breakers to be tested by actuating the relays which issue the tripping signal from the source.

All tripping matrix of low voltage circuit breakers to be tested by actuating the relays which issue the tripping signal from the source.

Intertripping signals tests between high and low voltage side of transformer.

Transformer mechanical relays tripping tests simulation, such as winding temperature, Buchholz relays, etc.

5.7.5 Load Test

Before energizing the transformer circuit bay, we connect the transformer bay to an empty busbar in the substation. With the acceleration of overcurrent relay to have a fast trip for any fault, (This means to reduce the time delay setting of the overcurrent relay) also to measure the voltage in secondary of the transformer bay circuit and the reference circuit in substation this means measure between R Phase of transformer circuit and R Phase of reference circuit it should read zero also for the other two phases Y and B also phase rotation should be checked before loading the transformer. This will be explained later in commissioning tests.

Chapter 6

Transmission Lines Theory Testing and Commissioning

6.1 INTRODUCTION

A transmission line is the electrical line that transfers power from one place to another. It can be over a short distance or long distance. Lines can be overhead lines or underground cables, and either AC or DC transmission.

6.2 OVERHEAD AC LINES

AC overhead transmission lines consist of conductors, earth wires, insulators and towers. They are used for voltages in the range of 66−800 kV. This high level of voltage is to reduce power losses during power transference through the line.

Power conductors are commonly made from aluminum with a stranded steel core (aluminum conductor steel reinforced). In higher voltages it is bundled to form one phase of the overhead line to reduce the corona effect (electric field) and radio interference. The earth wire does not carry any current and is made from high-strength steel or steel and aluminum. Insulators isolate the live phases from the tower body which represent the earth. Towers are supporting conductors.

6.2.1 Electrical Characteristics of AC Transmission Lines

The characteristics of these transmission lines consist of resistance (R), inductance (L), and capacitance (C) uniformly distributed over the length of the line. Fig. 6.1 shows the common representation of transmission line in π connection.

Overhead transmission has a loading capability limit depending on its length and system voltage level as shown in Fig. 6.2.

The load carrying capability of cable is also limited due to the charging current. The longer the cable, the higher capacitance, the higher charging current, and the higher cost of using compensation equipment to compensate for the reactive power supplied by this capacitance. The line carrying capability can be improved by using a series capacitance compensation for long

Practical Power System and Protective Relays Commissioning.
DOI: https://doi.org/10.1016/B978-0-12-816858-5.00006-X
69

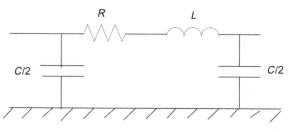

π-Representation of transmission line

R = line resistance in Ω,
L = line induction in Henrys
X = line capacitance in F
XL = inductive reactance = $2\pi.F.L$ Ω
XC = capacitive reactance = $1/(2\pi.F.C)$ Ω
Where F = power frequency = 50 or 60 HZ

FIGURE 6.1   Common representation of overhead transmission line.

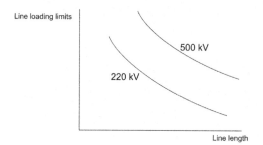

FIGURE 6.2   Loading of overhead transmission lines limit with line length and voltage level.

lines. The Ferranti effect is also associated with long lines; this means the reactive power in transmission line due to the line capacitance in a no-load condition; the charging current of this capacitance causes the rise of the voltage at the open receiving end of the line which can cause damage of breakers, surge arresters, current and voltage transformers, and transformers at the receiving end of the line. This problem needs to be compensated by using shunt reactors. The Ferranti effect can also exist in cables and requires shunt reactors to compensate. Refer to Fig. 6.3.

6.3   UNDERGROUND CABLES

Underground cables are more expensive than overhead lines; they also have a high capacitance that absorbs charging current that must be supplied by generators of the system, which is an additional cost. Underground cables are usually used in very populated areas, or underwater.

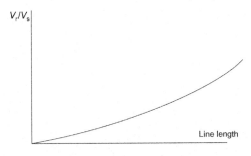

Where :

V_r = receiving end voltage of transmission line and

V_s = sending end voltage of transmission line

FIGURE 6.3 Ferranti effect voltage rise at line receiving end on open circuit condition.

Several types of high-voltage underground cables are used, for example, oil field type, cross-linked polyethylene (XLPE) cables, and SF6 Cables. There are some factors that affect the selection of cable types as follows:

1. Load characteristic of the cable.
2. Voltage rating of the cable which includes:
   a. Operating voltage V (line to earth) voltage during normal operation of the cable.
   b. Insulation voltage V_0 (Line to Line) voltage of the cable.
   c. In earthed networks $V = \sqrt{3} \times V_0$.
   d. In nonearthed networks, V should not exceed $1.7 \times \sqrt{3} \times V_0$.
3. Cable cross-section area should be chosen based on the following factors:
   a. Current carrying capacity.
   b. The value and the duration of overloading.
   c. The permissible voltage drop through the cable.
   d. Shortcircuit currents.
   e. Requirements of cable termination sealing end.

There is a classification for cable types based upon voltage levels as follows:

1. Low-voltage cables: up to 1000 V
2. Medium-voltage cables: 1−33 kV
3. High-voltage cables: 66−400 kV
4. Extra-high-voltage cables:400−750 kV

There is also a classification for cable types based on insulation type as follows:

1. Cables insulated with oil-filled paper.
2. Polyvinyl chloride (PVC) cables.
3. XLPE cables.

One can also classify cables based upon their construction:

- Conductor: aluminum or copper.
- Conductor shield.
- Insulation.
- Insulation shield.
- Filling material: plastic or PVC.
- Armor.
- Jacket or sheath.

6.3.1 High-Voltage Cables

Most high-voltage cables are:

- Paper-insulated oil-filled cables up to 500 kV AC. This type of cable is not good for the environment and requires maintenance.
- XLPE cables up to 525 kV AC. These cables can be in service for more than 20 years, with reliable good insulation, being maintenance free and good for the environment.

6.3.2 High-Voltage Cables: End Terminations

The ends of high-voltage cables can be terminated in a variety of ways:

Outdoor sealing end.
Transformer sealing end.
SF6 Switchgear sealing end.

6.4 TESTING AND COMMISSIONING OF EXTRA-HIGH-VOLTAGE AND HIGH-VOLTAGE CABLES AT SITE

6.4.1 Continuity and Phasing Checks

The continuity test is done after completion of all joints and termination installations to ensure that there are no broken conductors in any phase. This test is undertaken using a Megger tester, connecting one lead of the Megger to one phase of the cable and the other lead to the other phase. Each phase is tested as shown in Fig. 6.4.

The phasing test is done after completion of all joints and termination installations to ensure the cable phase is the same at both ends of the cable. This test is undertaken using a Megger tester, connecting one lead of the Megger to one phase and the other end of the cable phase to the earth. The other end of Megger tester is to be connected to the earth. By this test we identify the cable phases R, Y, and B at both sides of the cable as shown in Fig. 6.5.

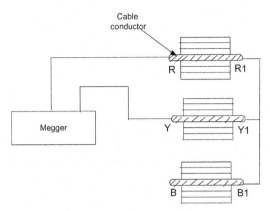

FIGURE 6.4 Cable continuity test.

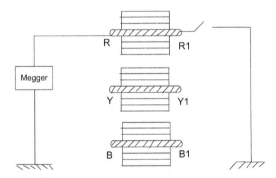

FIGURE 6.5 Cable phasing test.

6.4.2 DC Contact Resistance of Cable Phase Conductors

This test is undertaken using a microohmmeter that injects 100 DC Amp to measure the resistance as shown in Fig. 6.6.

One lead of the tester is connected to the first end of cable phase, and the other end of the tester is connected to the other end of cable phase. The same procedure is repeated for the other two phases of the cable.

The measured value should be converted to 20°C temperature reading by using the following equation:

$$\text{DC resistance reading at} 20°C = \frac{\text{DC resistance reading at} (t) \text{degree}}{(1 + \propto (t - 20)) \times L}$$

where L is the cable length in km, t is the temperature (°C) at which the test is done, α is the temperature coefficient for copper $= 0.0039$.

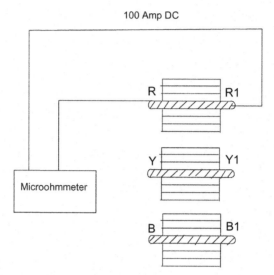

FIGURE 6.6   Cable conductor DC resistance test.

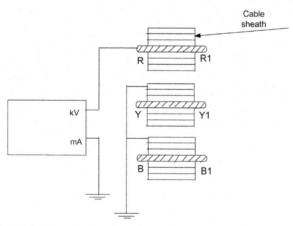

FIGURE 6.7   DC insulation sheath test.

6.4.3   DC Insulation Sheath Test

This test is done by using 10 kV for 1 minute. During this test, all links in the link boxes of the cable should be disconnected from the earth. The leakage current is measured as shown in Fig. 6.7.

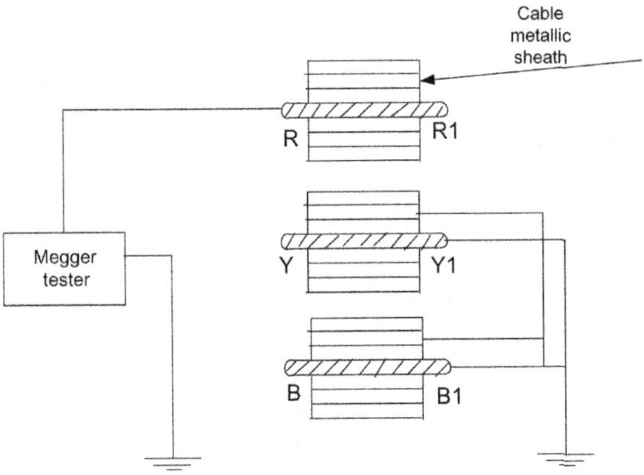

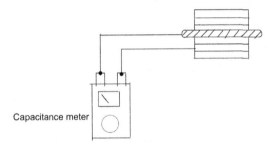

FIGURE 6.8   Cable insulation resistance test.

FIGURE 6.9   Cable capacitance measurement test instrument.

6.4.4   Cable Insulation Resistance Test

The cable insulation test is undertaken using a Megger tester of 5 kV DC for 1 minute. before and after the high voltage test of the cable. The connection will be as shown in Fig. 6.8.

The above procedures are repeated for the other two phases of the cable.

6.4.5   Cable Capacitance Test

This test is done using a capacitor meter as shown in Fig. 6.9.

The first lead of the capacitor meter is connected to the cable phase conductor and the other lead is connected to the cable metallic sheet and measured in μF.

6.4.6 Verification of Cross Bonding Test of Metallic Sheath of Cable

Cross-bonding is used to reduce the induced currents and eddy currents in the metallic sheath. This cross-bonding is made at link boxes. Transposition between phases is also made through the cable length. The test is undertaken by injecting an AC current of 100 A at the three phases of the cable with the other end of phases shorted. Then a measurement of the current passing through the cable sheath is made. This current should not exceed 3% of the injected current as shown in Fig. 6.10.

6.4.7 Measurements of Cable Positive Sequence Impedance

This test can be done by injecting a three-phase power supply on the cable with shorting the three-phase conductors of the cable at the remote end of the cable as shown in Fig. 6.11.where $Z_1 = Z_2 = V/I$ and Z_1 is the positive sequence impedance of the cable, Z_2 is the negative sequence impedance of the cable

$$R_1 = \frac{V}{I} \cos\Phi, \ X_1 = \frac{V}{I} \sin\Phi$$

6.4.8 Measurements of Cable Zero Sequence Impedance

This test can be done by using single phase power supply as shown in Fig. 6.12.where $Z_0 = 3 \ V/I$ and Z_0 is the Zero sequence impedance of the cable sequence impedance of

$$R_0 = 3\frac{V}{I} \cos\Phi, \ X_0 = 3\frac{V}{I} \sin\Phi$$

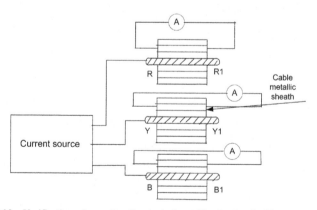

FIGURE 6.10 Verification of cross-bonding test of metallic sheath of cable.

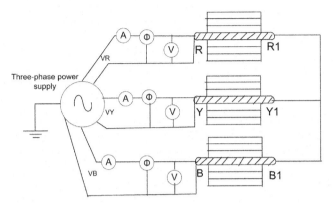

FIGURE 6.11   Measurements of cable positive sequence impedance.

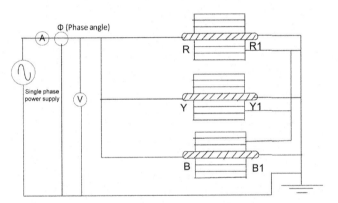

FIGURE 6.12   Measurements of cable zero sequence impedance.

6.4.9   Sheath Voltage Limit Test

At the cross-bonding link boxes, a nonlinear resistance is used to reduce the induced voltage in the cable sheath during a fault condition to test the insulation of the sheath voltage limit (SVL). Leave the three terminals of SVL connected to earth and the other three terminals are then able to measure each free terminal of the SVL. Record using the Megger tester after adjusting to 1 kV DC for 1 minute. Then discharge the induced charge at all the free terminals after each measurement. Restore the connection after test, as shown in Fig. 6.13.

The insulation value of the SVL should be more than 1 GΩ at 1 kV.

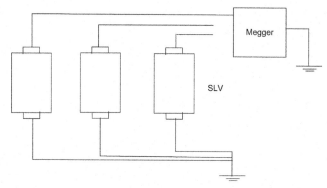

FIGURE 6.13 Sheath voltage limit insulation test.

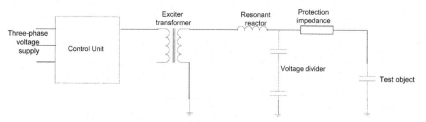

FIGURE 6.14 High voltage resonant AC test for cables.

6.4.10 High-Voltage Resonant AC Test for 132 kV Cables and Above

The following steps should ensure this test is carried out in a safe way:

1. Jumper-lead all SVLs of the cable system under test and restore the case after test.
2. Before and after the high-voltage test, measure and register the cable insulation resistance.
3. AC 260 kV at 20−300 Hz frequency as shown in Fig. 6.14 shall be applied for 60 minutes through a frequency-tuned series resonant test set between the conductor and the metallic sheath of each individual XLPE cable.
4. All cable conductors and shields not on test shall be grounded.

   Refer to Fig. 6.14.

6.5 HIGH VOLTAGE DC POWER SYSTEMS TRANSMISSION

6.5.1 Introduction

A DC transmission consists of converters, inverters, lines with two conductors and is used economically for long transmission lines.

In this mode of transmission a three-phase 50 or 60 Hz voltage and currents are converted to DC voltage. Currents transmitted on a bipolar line consist of two conductors at the end of the line converted back to AC voltages and currents by inverters.

The advantage of this system is that it can connect between two different frequency AC systems. It can also be used in long underground cables or underwater submarine cables, and is very good for long distance overhead transmission. It is used for transmission lines over 640 km and over 32 km for underground cables.

A simple DC transmission system consists of transformers at both ends of the line, a converter from AC to DC, and an inverter from DC to AC again at the end of the line. Filters for harmonics at both ends of the line are also used (see Fig. 6.15).

This system of transmission is used to transfer power for long distances. It has more advantages than the AC transmission such as:

1. No charging current in the system.
2. No skin effect on conductors.
3. No compensation equipment for reactive power is required.
4. No synchronization is required to tie between two systems.
5. Even with a tie facility between two AC systems, the two AC systems have different frequencies.
6. Shortcircuit current is less.
7. The fault will not transfer between the two AC systems tied by a DC link.
8. The power losses and interfaces with radio and TV waves is less in DC transmission than AC transmission.
9. The insulation level in DC transmission is less than AC transmission, reducing costs.

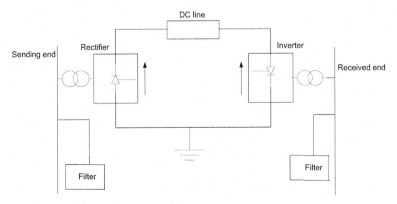

FIGURE 6.15   Simple DC transmission line system.

10. No stability problems in DC transmission as there are no constraints on line length.

11. This system is economic for long transmission lines.

However, there are still some limitations or disadvantages of using DC transmission:

1. High costs of inverters and converters station at the line ends.

2. Harmonics are generated and these harmonics can interfere with the telecommunication equipment, which lead to using filters which add to the cost of a DC transmission system.

3. There is no zero crossing point in DC voltage , which requires advanced technology to produce the DC circuit breakers. Which means that in some cases the DC transmissions are done without the use of DC circuit breakers.

6.5.2 Construction of DC Transmission

DC transmission can be configured in monopolar, bipolar, or homopolar links. These are discussed below.

6.5.2.1 Monopolar links

Uses one conductor and the return path is provided by ground as shown in Fig. 6.16.

6.5.2.2 Bipolar links

Each terminal has two converters of equal rated voltage, connected in series on the DC side. Junctions between the converters are grounded. The

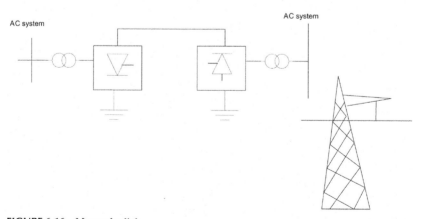

FIGURE 6.16 Monopolar link.

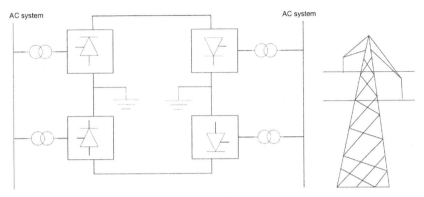

FIGURE 6.17   Bipolar links.

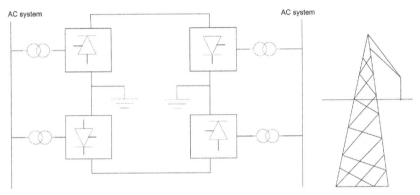

FIGURE 6.18   Homopolar links.

advantage of this configuration is that if one pole is faulty the other pole will operate with half load transfer (see Fig. 6.17).

6.5.2.3   Homopolar links

Homopolar links have two conductors with the same negative polarity. The return path is through the ground (see Fig. 6.18).

6.5.3   DC Transmission Components

This system consists of converter transformers at the ends, filter, and a DC line as shown in Fig. 6.19.

In more detail the system consists of:

1. A converter that converts AC to DC or DC to AC and consists of valves connected in a 6-pulse or 12-pulse arrangement and transformers.

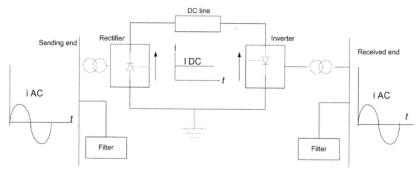

FIGURE 6.19 DC transmission components.

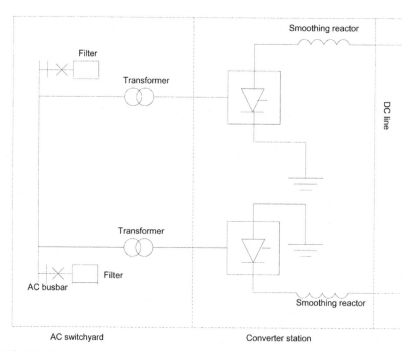

FIGURE 6.20 High-voltage DC system components.

2. Air core smoothing reactors, used in series with each pole to decrease harmonics in DC line currents and voltages.
3. Harmonic filter, used to reduce harmonics generated by converters in current and voltage. These harmonics causes heating of system components and interfere with telecommunication equipment.
4. Control system to the DC link, which provides power control and frequency control and DC protection.
5. AC switchgear. Refer to Fig. 6.20.

6.5.4 High-Voltage DC Protection System

The protection system is divided into two parts; one for the DC components and the other for AC components:

1. DC protection system:
   a. Converter protection.
   b. Pole protection.
   c. DC switchyard protection.
   These functions can be performed by the control system of the DC link.
2. AC protection systems:
   a. Converter transformer AC protections.
   b. AC bus protections.
   c. AC filter and reactor protections that includes mechanical protections for the reactor.

Chapter 7

Circuit Breakers Theory Testing and Commissioning

7.1 INTRODUCTION

Circuit breakers vary from simple molded case breakers used in 120-V lighting to complex units used at 750 kV. The major factors in the design of CB include the method of extinction of fault currents and the speed of the operating mechanism, arc-interrupting medium, and insulation system.

CB functions are to:

- open and close the electrical circuit manually and automatically and
- trip in case of short circuit to isolate the faulty part in the system.

CB ratings include the following:

1. Voltage ratings
   a. Nominal kV rating
   b. Maximum voltage at which the breaker may be operated;
2. Insulation level withstand voltage Root Mean Square Value (RMS) kV at 50 or 60 Hz;
3. Current ratings
   a. Continuous 50- or 60-Hz rating in amperes—current that the breaker carries in normal conditions without exceeding the heating limits
   b. Short-time rating of approximately 1.6 of the rated interrupting current
      a. Momentary rating up to 1 second
      b. Four-second rating;
4. Interrupting rating is highest the current breaker should interrupt at normal voltage, in general it is about 16 times the rated continuous current of the breaker
   a. Three-phase rated MVA—interrupting capacity
   b. Amperes at rated voltage
   c. Maximum amperes;
5. Operating time in cycles
   a. Closing time is the interval between the energizing of the closing circuit and the making of the arcing contacts of the breaker

Practical Power System and Protective Relays Commissioning.
DOI: https://doi.org/10.1016/B978-0-12-816858-5.00007-1

b. Contact parting time is the interval between the energizing of the trip coil and the parting of the primary arcing contacts of the breaker

c. Arcing time is the interval between the parting of primary arcing contacts and the extinction of the arc on the primary contacts of the breaker

d. Tripping time—interrupting time is the interval between the energizing of the trip coil and the extinction of the arc on the primary contacts of the breaker

e. Reclosing time is the interval between the energizing of the trip coil when the breaker is in the close position and the making of primary arcing contacts on the reclosing stroke

f. Trip-free time is the interval between the touching of the main contacts and the parting of the main contacts when the breaker has attempted a close into unknown fault condition

g. Breaker duty cycle: each breaker has a sequence of operation cycles depending on the breaker mechanism reset time between the opening and closing operation and between the closing and opening operation

h. Closing (C)

i. Opening (O)

j. Operating sequence or duty cycle of CB.

   O—0.3 s—CO—3 min—CO.

7.2 PRINCIPLE OF ARC INTERRUPTION

The CB should provide an insulating medium that is sufficient to prevent current from continuing to flow during current interruption. The circuit is usually opened by drawing out an arc between contacts until the arc can no longer be sustained and is extinguished as shown in Fig. 7.1.

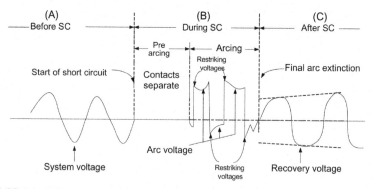

FIGURE 7.1 Voltage waveform of the circuit breaker interruption process. (A) Before short circuit (SC); (B) during SC; and (C) after SC.

The extinguishing of the arc is done at a current of zero and a dielectric is injected between the contacts faster than the recovery voltage can build up. This process is the same regardless of the dielectric medium used.

Fig. 7.1 shows only the voltage waveform from a normal operating condition, through a short circuit, and then back to the recovery voltage. Fig. 7.1A shows the system operating under normal conditions. At the beginning Fig. 7.1B shows a short circuit has occurred, which will cause the voltage to theoretically go to zero during the prearcing stage. As the contacts begin to separate, the short circuit current that was following just prior to the contacts separating continues to follow and establish an arc. The arc interruption will depend on the type of insulating media (air, oil, vacuum, SF_6). During the arc stage (Fig. 7.1B) shows the arc is extinguished at each current zero, due to ionization of the insulating media gap the arc will once again begin to form due to the system voltage across the contact gap. This condition of current flow again causes the restrike voltage peaks to occur. Then, as shown in Fig. 7.1C, after the contact has separated the short circuit should be extinguished and the recovery voltage appears.

Fig. 7.2 shows the break downs in the current zero window of the restrike condition.

Fig. 7.3 shows the recovery condition in the current zero window.

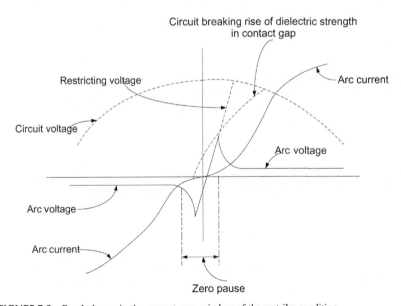

FIGURE 7.2   Break downs in the current zero window of the restrike condition.

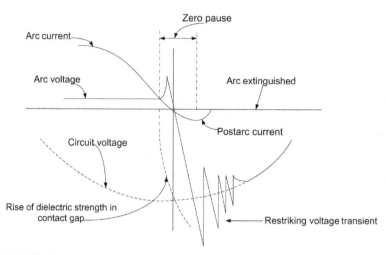

FIGURE 7.3 The recovery condition in the current zero window.

7.3 CIRCUIT BREAKER TYPES BASED ON INSULATING MEDIUM

7.3.1 Air Circuit Breaker

An air CB employs air as the interrupting insulation medium. The air CB poles consist of a set of main contacts and a set of arcing contacts. The arcing contacts open after the main contacts and the arc is drawn on them. A puff of air from a puffer is used to aid in blowing the arc deeper into the throat of the arc chute.

A blast of air trained on the arc from a reservoir of compressed air enables a heavier arc to be extinguished in an air blast CB.

Refer to Fig. 7.4 for an air CB contact assembly.

7.3.2 Oil Circuit Breaker

Oil CB have their contacts immersed in insulating oil, during Oil Circuit breaker opening the insulating medium- Oil- will cools the arc, the heat and chemical reaction create hot gases which expand rapidly, blowing the arc out. There are two types of oil CB: the original type was the bulk oil CB and the current one is the minimum oil CB. Refer to Fig. 7.5 for a schematic of a bulk oil CB.

Fig. 7.6 illustrates a schematic of a minimum oil CB.

7.3.3 Vacuum Circuit Breaker

In this breaker the isolating medium is a vacuum in a chamber in which the air has been evacuated as illustrated in Fig. 7.7.

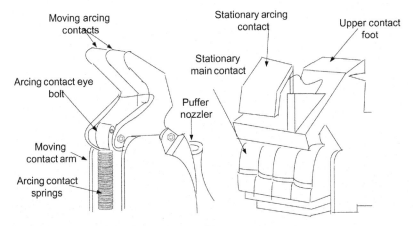

FIGURE 7.4 Air circuit breaker contact assembly.

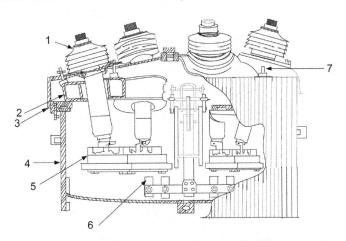

1- Bushing   2- Bushing current transformer   3- BCT support plate

4- Tank   5- Interrupter   6- Contact rod   7- Oil gage

FIGURE 7.5 Schematic of a bulk oil circuit breaker.

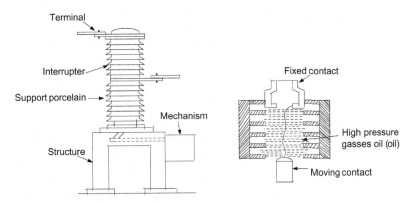

FIGURE 7.6 Schematic of a minimum oil circuit breaker.

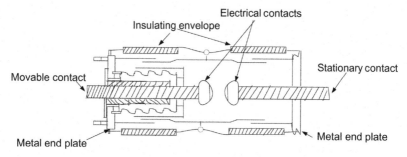

FIGURE 7.7   Schematic of a vacuum circuit breaker.

As the arc in an air CB is an electric conductor made up of ionized air particles, the arc is extremely difficult to form in a vacuum. This type of breaker has limited interrupting ratings compared to oil CB.

7.3.4   SF₆ Circuit Breaker

SF_6 (sulfur hexafluoride) is an excellent gas for insulation in high-voltage equipment due to its excellent dielectric and arc-quenching properties in a gas insulated system (GIS), where it is used in CB, disconnect switches, bus ducts, current and voltage transformers, etc.

SF_6 is an industrial product developed through the chemical combination of fluorine and sulfur. The resulting product is purified, liquefied by compression, and stored in steel cylinders.

7.3.4.1   Properties of SF₆ Gas

SF_6 gas has the following characteristic:

- odorless,
- colorless,
- nontoxic,
- nonflammable,
- heavier than air, and
- good characteristic for electrical arc quenching.

Normally it is stored in pressurized cylinders.

7.3.4.2   Contaminants

Contaminates, such as water, air, and oil, come as a result of poor vacuum processing of the gas and reduces its electrical dielectric.

7.3.4.3   Related Standards

IEC Standard 376 recommends the following limits:

CF4 maximum of 500 ppm,
oxygen and nitrogen (air) of 500 ppm,

water of maximum 15 ppm,
acidity of 0.3 ppm maximum, and
hydrolysable fluorides maximum of 1.0 ppm.

7.3.4.4 Precautions

1. Ventilation
   As SF_6 is heavier than air, it replaces air in the room and can cause asphyxiation.
2. Toxic arc by-products
   For any fault inside SF_6 CB, care should be taken for the products of SF_6 in arc chamber of the CB, and people should wear protective clothes and protective equipment.
3. SF_6 gas contamination by compressor oil.
   Compressors are the source of oil contamination; oil test unit is required to test the oil contamination.

7.3.4.5 Testing for Contamination

1. Moisture content
   Moisture is measured using dew point measuring instrument where readings are directly in ppm moisture.
2. Oxygen content
   Analytical portable trace oxygen analyzer will be used for measurement, and the measurement will be in $1-1000$ ppm by weight.
3. Gaseous by-products
   During internal faults, tests have to be done for the gas products.
4. Gauge calibration
   4.1. Density monitor checks
      The density monitor should be recalibrated using variable compressed air.
   4.2. Pressure and gauge checks
      Gauge recalibration should be performed. Refer to Fig. 7.8.
      a. This figure shows monograph for dew points and ppm as a function of pressure, for example, at a pressure of 3 bars.
      b. $-20°C$ will give a moisture content of 350 ppmv on line of monograph.

7.4 CIRCUIT BREAKER OPERATING MECHANISM

1. A stored energy mechanism uses large springs.
2. A penumatically operated mechanism employs compressed air in the operation of the CB.
3. A hydraulic operated mechanism employs compressed oil.

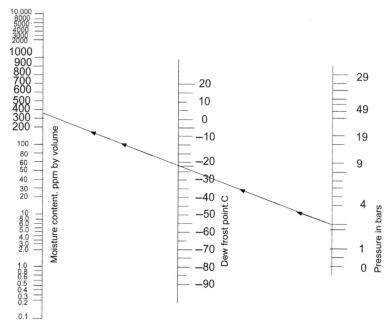

FIGURE 7.8 Monograph for dew points and ppm as function of pressure.

7.5 CIRCUIT BREAKER CONTROLS

7.5.1 Electrical Control

Before the introduction to the trip and closing circuit of CB, some circuits are described.

1. Cut-off circuit

    The closing coil has a short-time rating of 15 seconds, so it is necessary to disconnect the power on this closing coil after the closing operation is completed; this is called the cut-off circuit.

2. Seal in "self-holding" circuit

    When the closing operation is started a self-holding process is initiated for the closing circuit, after the operator presses the closing pushbutton a closing relay is energized and closes its self-holding contact to ensure the power is continuing to the closing relay until the end of the closing stroke of the CB and this power is cut only by an auxiliary contact of the CB.

3. Antipumping circuit

    When there is a fault in primary system occurs,the protective relays will issue a trip signal for the breaker during this moment the operator initiating a close order (manual close) then a repeated operation of the trip and close cycle will occur, which can damage the CB, therefore there is a relay to open the closing circuit during the trip signal only, called the antipumping relay. This can happen also if the contacts of pushbutton jammed in closed position.

4. Trip-free CB feature

   If the breaker is able to receive a trip signal and to trip before a closing operation is completed then this breaker is called a trip-free CB.

5. Antislamming circuit

   When a CB is in a closed position and receives a close order this can cause slamming of the CB mechanism. The antislaming circuit open the d.c of the closing circuit of the CB when the CB is in the closed position.

6. CB pole discrepancy tripping circuit

   If for any reason, mechanical or electrical, the three phases of the CB are not fully closed or all fully open at a given time, then a three-phase trip is issued for the CB phases.

7. CB low hydraulic oil pressure alarm and lockout circuit

   The purpose of this circuit is to give an alarm when the breaker has a low hydraulic system oil pressure and indicates that it is necessary to operate the pump to raise the oil pressure. If the oil pressure continues to decease below a certain level then a block for trip and close is issued by this circuit to prevent the CB from operation.

8. CB hydraulic pump fault alarm circuit

9. SF_6 CB low SF_6 gas pressure alarm and lockout circuit

   If the pressure of the insulating medium gas (SF_6) is low then this circuit issues an alarm. If the pressure continues to decrease below a certain level then this circuit issues a three-phase trip signal for the CB and locks out its operation closing and opening, and sends a trip signal to the remote end of the high-voltage transmission line to trip the remote end breaker.

   Refer to Fig. 7.9 for CB closing circuit.

As shown in Fig. 7.9, the operator can close the breaker from three different locations:

1. Local, at the switchgear;
2. Remote, at the control panel in the control room or control desk computer in the control room in new substations;
3. From National Control Center (NCC) (Dispatch Center) to the master station.

Only one close order can be issued at a time—this order is done through manual synchronizing or through auto synchronizing relay.

It is shown from the circuit that autoreclosing can also be done but through auto synchronizing relay.

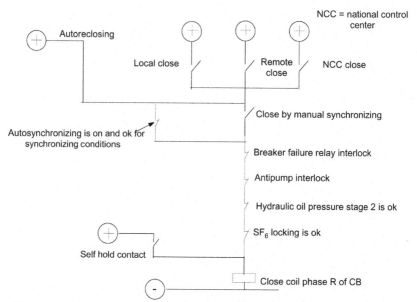

Simple circuit breaker close circuit for hydraulic mechanism SF$_6$ circuit breaker

FIGURE 7.9 Closing circuit of circuit breaker.

Also, it can be seen that the following conditions should be correct in order to complete the closing order:

1. SF$_6$ pressure is correct.
2. Oil pressure for mechanism is correct.
3. There is no antipumping condition during the closing condition.
4. Breaker failure relay is not operated.

Refer Fig. 7.10 for the trip circuit of a CB.

As shown in the circuit in Fig. 7.10, the tripping can be done through protection relay contact, or from the NCC, through pole discrepancy circuit, or by a manual trip by the operator.

The trip should be done when the oil pressure of the hydraulic mechanism is correct and the pressure of the insulating medium SF$_6$ is correct also.

7.6 CIRCUIT BREAKER TESTING

7.6.1 Contact Resistance Testing

This test is to confirm the resistance of the main contacts.

A 100 A DC current is injected through the main contact by keeping the CB closed. The voltage drop across the contact is measured and the

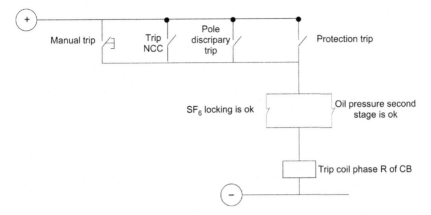

Simple circuit breaker trip circuit for hydraulic mechanism SF$_6$ circuit breaker

FIGURE 7.10   Trip circuit of circuit breaker.

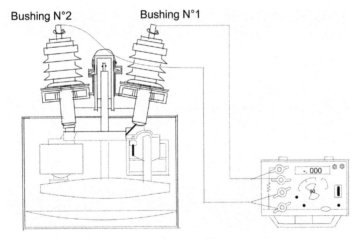

FIGURE 7.11   Microohmmeter test connection for a CB contact resistance test.

resistance calculated. In many instruments will indicate a reading directly in micro-ohms as shown in Fig. 7.11.

The obtained values should be compared with factory test reports or the manufacturer's claimed values.

7.6.2   Insulation Resistance Testing

The test voltage should be between the phase to earth and across the poles, by measuring each phase to earth (or body) and across the pole for each phase.

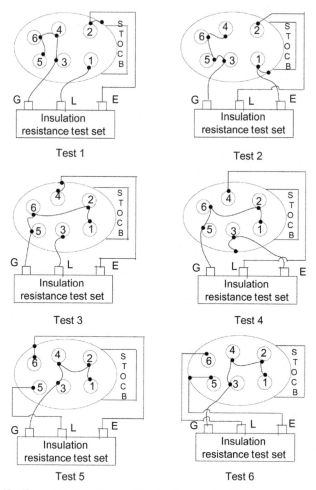

FIGURE 7.12 Test connections for circuit breaker in open condition.

The applied test voltage limits shall be in the range of 1000−2500 V DC for CB's up to 5 kV (AC) and up to 5000 V DC for CB's above 5 kV (AC).

Test connections are shown in Figs. 7.12 and 7.13 for an outdoor oil CB with six bushings. Fig. 7.12 shows the test connections for a CB in open condition and Fig. 7.13 shows the test connections for a CB in closed condition, where STOCB is a single tank oil CB; G is the guard, L is the phase, and E is the earth.

7.6.3 High-Voltage Test

This test is done during the switchgear high-voltage test. The AC high-potential test will stress the CB insulation similarly to the stresses which

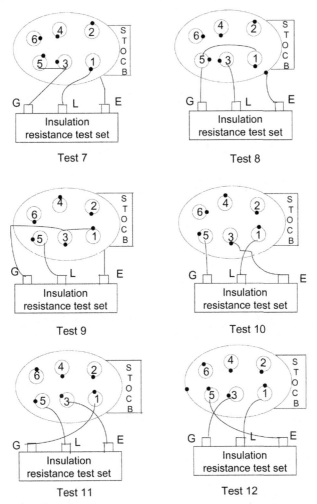

FIGURE 7.13 Test connections for circuit breaker in a closed condition.

occur under abnormal voltage conditions during system operations. The high-potential test should be the final test performed after all repairs have been made and a successful insulation resistance test has been carried out. The maintenance test voltage should be 75% of the factory test voltage and the time duration should be 1 minute. For example, for 13.8 kV the factory test voltage will be at 36 kV and the maintenance test voltage will be at 27 kV.

7.6.4   Circuit Breaker Timing Test

This test is performed to verify the open and closing times of CB contacts. It is done using a CB timing tester to measure the closing time and tripping

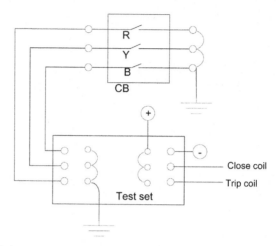

CB timing measurement connections using test set shown above

FIGURE 7.14 Time measurement connection to measure the operating times of circuit breakers.

time—this device measures and records the time and timing diagram. The test is done at 100% of voltage substation DC voltage and 80% of voltage substation DC voltage, as well as at 60% of voltage substation DC voltage and at 80% there should be no any pole discrepancy in three-phase closing and three-phase tripping.

The obtained close/open time should be compared with the manufacturer's reference values or factory test results. Refer to Fig. 7.14 for CB timing tester connection to measure the operating times of CB.

7.6.5 Reduced Voltage Test

This test is carried out to confirm the operation of the closing coil of the circuit breaker when reducing the control dc voltage of the circuit breaker. A reduced voltage of 80% of the rated voltage for the closing coil and 60% of the rated voltage for the trip coil at these two values of the voltage the breaker should be able to operate (close and open).

Chapter 8

Air Insulated System Substations Theory and Testing

8.1 INTRODUCTION

A high-voltage air insulated system (AIS) consists of the following items:

1. Primary plant
   a. Heavy current busbars
   b. Switchgear, transformers, etc.
2. Secondary equipment
   a. Instrument transformers
   b. Protection, control, and information systems
3. Infrastructure
   a. Civil foundations and structures

Auxiliary systems.As shown in Fig. 8.1 this system uses air as insulating medium. This has restrictions in the minimum clearance distance required in order to be safe. For example, for a 220 kV system. the clearance distance is 178 cm for phase to earth and 225 cm between phases. For a 66 kV system the clearance distance is 63 cm for phase to earth and 77 cm between phases.

Figs. 8.2−8.6 show the components for AIS. Testing of these components will be discussed in the following chapters.

Practical Power System and Protective Relays Commissioning.
DOI: https://doi.org/10.1016/B978-0-12-816858-5.00008-3

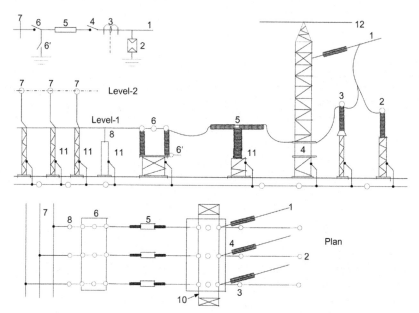

FIGURE 8.1 Air insulated switchgear line construction. 1: line; 2: surge arrester; 3: current transformer; 4: isolator line side; 5: circuit breaker; 6: isolator (disconnector); 6′: earthing switch; 7: busbar tubular rigid; 8: postinsulator support; 9: strain insulator for 1; 10: gantry (beam); 11: support structure; 12: shielding conductor (ACSR).

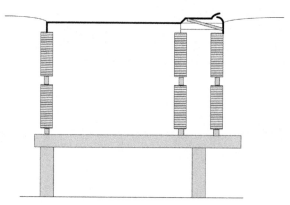

FIGURE 8.2 Horizontally-upright mounted disconnect switch.

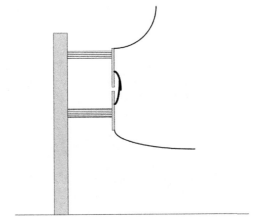

FIGURE 8.3   Vertically-mounted disconnect switch.

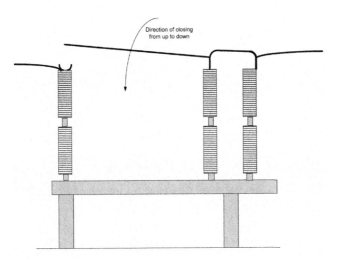

FIGURE 8.4   Vertical break disconnect switch.

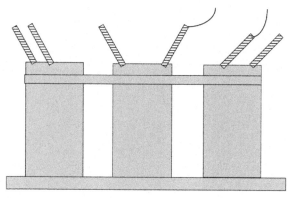

FIGURE 8.5   Three tank bulk oil circuit breaker.

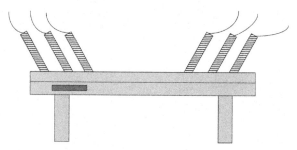

FIGURE 8.6   SF6 gas dead tank circuit breaker.

8.2   TESTING OF THE AIS-COMPONENTS

Explained in details in previous Chapter 3, Introduction to Testing and Commissioning of Power System.

Chapter 9

Surge Arresters Theory Testing and Commissioning

9.1 INTRODUCTION

A surge arrester protects system equipment such as transformers and transmission lines from excessive voltage and/or overvoltage caused by lightning or switching surges. The old design was a gap in series with a nonlinear resistor all enclosed in porcelain cover for protection. The resistance is very low at high voltage and high for low voltages; during high voltages the gap sparks over to protect the equipment, as shown in Fig. 9.1.

A new design uses zinc oxide for surge arresters, that have a nonlinear voltage current characteristic as shown in Fig. 9.2.

In normal voltage the arrester is open circuited, however when surges in voltage occur, the arrester is short circuited to pass the current and protect the equipment. Arresters are applied both at the transformer side(s) and also the circuit breaker side(s) to protect the circuit breakers when it is in the open position and during its opening.

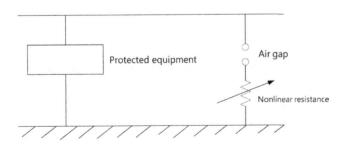

FIGURE 9.1 A simple surge arrester.

Practical Power System and Protective Relays Commissioning.
DOI: https://doi.org/10.1016/B978-0-12-816858-5.00009-5

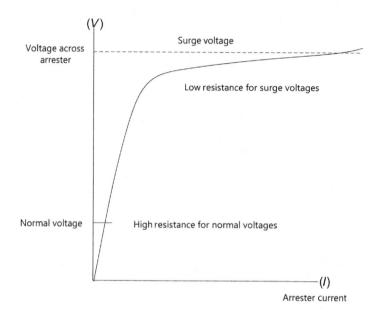

V-I characteristic of zinc oxide surge arrester

FIGURE 9.2 Zinc oxide surge arrester.

9.2 TESTING OF SURGE ARRESTERS

Leakage current measurements are shown in Fig. 9.3. I_t (Amp) is calculated by:

$$I_t = I_c + I_r$$

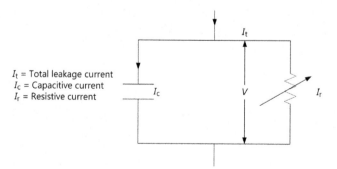

FIGURE 9.3 Leakage current measurements.

Chapter 10

Traditional and Electronic Current Transformers Theory Testing and Commissioning

10.1 INTRODUCTION

A current transformer (CT) is used to transfer the primary current to a secondary current proportional with the primary current. The secondary current is suitable for the measuring and protection devices.

Fig. 10.1 shows the construction, components, and types of some CTs used in most substations throughout the world.

10.2 CURRENT TRANSFORMER EQUIVALENT CIRCUIT

A CT consists of an iron core with two windings. The primary winding measures the primary current and the second winding connected with the secondary load (also called a burden). The current in the primary circuit produces an alternating flux in the core and this flux produces an emf in the secondary circuit that produces the secondary current, see Fig. 10.2. Refer to Fig. 10.3 for a current phasor diagram for CTs.

As shown in Fig. 10.3, the primary current is the summation of the secondary current (I_s), this corresponds to primary side plus the excitation current (I_e) which has two components: I_m to magnetize the core and I_c for iron and hysteresis losses.

10.3 CURRENT TRANSFORMER MAGNETIZATION CURVE

Refer to Fig. 10.4 for the CT magnetization curve. The protection CT is working between the ankle point and the knee point and beyond, but the measuring CT usually operates in the region of the ankle point. The knee point voltage is defined as the point at which an increase in voltage by 10% will increase the magnetizing current (I_e) by 50%. Above this voltage, the CT will be saturated.

Practical Power System and Protective Relays Commissioning.
DOI: https://doi.org/10.1016/B978-0-12-816858-5.00010-1
© 2019 Elsevier Inc. All rights reserved.

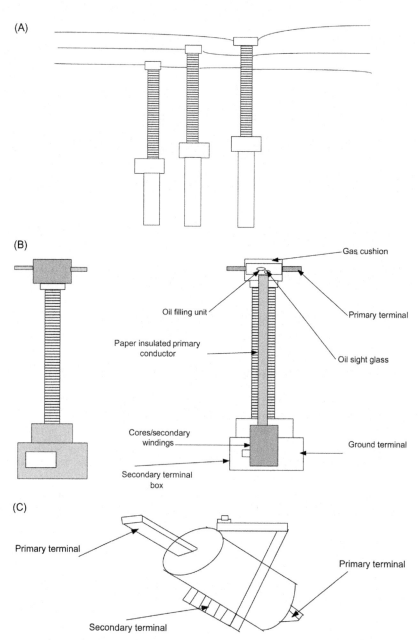

FIGURE 10.1 (A) Installed top-core current transformer (CT). (B) Oil-immersed hair-pin CT. (C) Bushing-type CT.

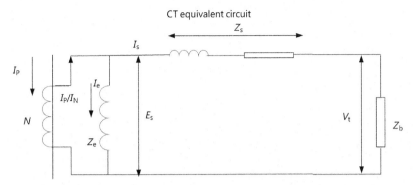

FIGURE 10.2   Current transformer equivalent circuit. I_p = primary rating of CT; N = CT ratio; Z_b = burden of relays in Ohms $(r + Jx)$; Z_s = CT secondary winding impedance in Ohms $(r + Jx)$; Z_e = secondary excitation impedance in Ohms $(r + Jx)$; I_e = secondary excitation current; I_s = secondary current; E_s = secondary excitation voltage; V_t = secondary terminal voltage across the CT terminals.

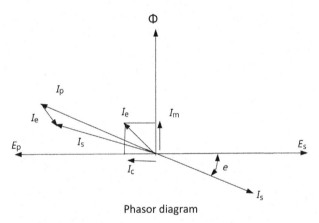

Phasor diagram

FIGURE   10.3   Current   transformer   current   phasor   diagram. E_p = primary   voltage; E_s = secondary   voltage; Φ = flux; I_C = iron   losses   (hysteresis   and   eddy   currents); I_m = magnetizing current; I_e = excitation current; I_p = primary current; I_s = secondary current.

Protection CTs work with high knee point voltage, but measuring devices work at the low knee point voltage as shown in Fig. 10.4.

In many applications, the core saturation of CTs will almost inevitable, and occur during the transient phase of a heavy short circuit. Once saturated, the equivalent circuit of the CT is as shown in Fig. 10.5A. The whole current is passing through the magnetizing path which is seen to be short circuited. Additionally, the ratio of the CT becomes disturbed due to the distortion in the secondary current (see Fig. 10.5B).

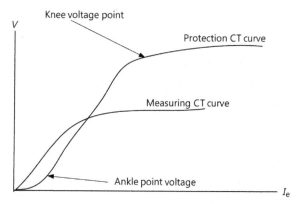

FIGURE 10.4 Current transformer (CT) knee point voltage in measuring and protection CTs.

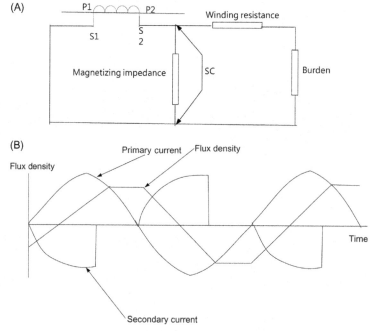

FIGURE 10.5 (A) Equivalent circuit of current transformer (CT) in saturation condition. (B) Secondary current of CT distortion at saturation.

At saturation the magnetizing impedance will be short circuited and the secondary current will pass through the magnetizing impedance. As the magnetizing impedance is nonlinear, the secondary current draft from the sine wave is as shown in Fig. 10.5B. The secondary current and voltage then collapses to zero until the next primary current zero is reached. This process is repeated each half cycle, resulting in a CT ratio that is disturbed after saturation.

10.4   CURRENT TRANSFORMER ACCURACY CLASSES

10.4.1   Metering Current Transformer Accuracy Classes

Refer to the tables below for the current error at rated frequency. Table 10.1 shows the limits of error for accuracy classes 0.1–5.

10.4.2   Protection Current Transformer Accuracy Classes

Class P CT are characterized by:

Turns ratio
Rated burden
CT burden is the load applied to the secondary of CT and expressed in terms of the impedance of the load and its resistance and reactance components. In practice it is expressed in volt-Amperes (VA)
Class 5P or 10P:
Accuracy limit factor (ALF), also called the saturation factor.
The CT is designed to maintain its ratio within specified limits up to certain value of primary current, expressed as multiple of its rated primary current. This multiple is called rated accuracy limit factor. For example, 15 VA 10P20 to convert VA and ALF into volts $V_K \approx \text{VA/IN} \times \text{ALF}$

For example, for a CT nominated as Class 5P20 means a CT composite error $\pm 5\%$ in secondary current at 20 times of primary current, as shown in Table 10.2.
Class X CTs (a typical curve is shown in Fig. 10.6) are characterized by:

Rated primary current
Turns ratio
Knee point voltage
Magnetization current at the knee point
Secondary resistance

TABLE 10.1 Limits of Error for Accuracy Classes 0.1–5

Class	Ratio (%) Error
0.1	0.1
0.2	0.2
0.5	0.5
1	1.0
3	3
5	5

TABLE 10.2 Accuracy Versus Composite Error Calculations

Accuracy Class	Composite (%) Error
5P	5
10P	10

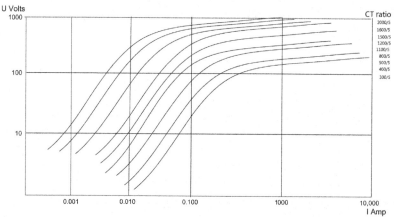

Typical example of CT excitation curve

FIGURE 10.6   Typical excitation curve of multiratio 2000/5 Amp current transformer class X, $V_K = 700$ V, 60 Hz, $R_{CT} = 0.85\ \Omega$.

Fig. 10.7 shows how to read the CT class for measuring and protection CTs.

Let us look at an example: a CT classified as 5P10.

The number 5 refers to the CT composite error in magnitude and angle.

The number 10 refers to its ALF.

This limit means a $\pm 5\%$ error at 10 times of primary current.

$$V_K = \frac{VA}{IN} \cdot ALF$$

Example:

CT VA = 15 VA 5P10

ALF = 10

$$V_K = \frac{VA}{IN}.ALF = \frac{15}{1}.10$$

$V_K = 150$ V, VA $= I^2.R$

Class $\times$ CT BS classified by:

CT ratio, I_m, R_{CT}, V_K, I_P

Measuring CT specification

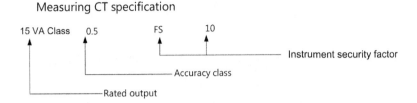

Protective CT specification

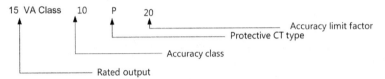

FIGURE 10.7   How to read the current transformer (CT) class for measuring and protection CTs.

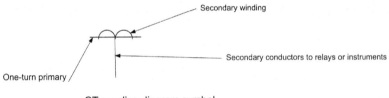

CT-one-line diagram symbol

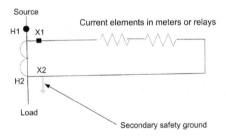

Polarity marks shown as dots

FIGURE 10.8   Current transformer in schematic diagrams.

10.4.3   Current Transformer Open Circuited Secondary Winding

The secondary winding of the CT should never be left open-circuited whilst the primary current continues to flow. In these circumstances, only the primary current is active and a peak high voltage appears at the secondary output terminals. This situation is dangerous and can result in the breakdown of the secondary circuit insulation of the CT (see also Fig. 10.8).

Equivalent CT accuracy ratings are shown in Table 10.3.

TABLE 10.3 Equivalent Current Transformer Accuracy Ratings

IEEE C 57.13	IEC 60044-1
C 100	25 VA 5P20
C 200	50 VA 5P20
C 400	100 VA 5P20
C 800	200 VA 5P20

10.5 TYPES OF CURRENT TRANSFORMERS

10.5.1 Wound Primary Type Current Transformer

This CT has conventional windings (primary and secondary) formed of copper wire wound round a core. It is used for auxiliary CTs and for many low or moderate ratio CTs used in switchgear of up to 11 kV rating.

10.5.2 Bar Primary Type Current Transformer (Resin-Embedded)

These CTs, also know as ring-wound transformers, have a ring-shaped core. The secondary winding occupies the whole perimeter of the core. Such CTs normally have a single primary conductor, sometimes permanently built-in and are provided with the necessary primary insulation.

10.5.3 Bushing-Type Current Transformers

Bushing CTs are usually less expensive than bar-primary and wound types. Bushing CTs are designed with an iron core encircling an insulating column through which the primary current lead connects to the bushing. The diameter of the iron core is large (to fit over large bushings) compared to other CTs, resulting in a large mean magnetic path length.

10.5.4 Air-Gapped Current Transformers

These are auxiliary CTs in which a small air gap is included in the core to produce a secondary voltage output proportional in magnitude to the current in the primary winding. This form of CT has been used as an auxiliary component of unit protection schemes in which the outputs into multiple secondary circuits must remain linear and proportioned to the widest practical range of input currents.

10.5.5   Transient Performance Current Transformers

IEC 60044-6 classifies these into types TPX, TPY, and TPZ. TPX is the closed iron-core CT while TPY and TPZ are CTs with air gaps in the core.

10.5.5.1   TPY Class

An antiremanence CT has a small gap in the core magnetic circuit, thus reducing the possible remanent flux from approximately 90% of saturation value to approximately 10%. This gap is quite small, for example 0.12 mm total, and so the excitation characteristic is not significantly changed by their presence. However, the resulting decrease in possible remanent core flux confines any subsequent DC flux excursion, resulting from primary current asymmetry, within the core saturation limits. Errors in current transformation are therefore significantly reduced by these CTs.

10.5.5.2   TPZ Class

Linear CTs constitute an even more radical departure from the normal solid core CT. It incorporates a large air gap, for example, 7.5−10 mm. As its name implies, the magnetic behavior tends to be linear by the inclusion of this gap in the magnetic circuit. However, the purpose of introducing more reluctance into the magnetic circuit is to reduce the value of the magnetizing reactance. This in turn reduces the secondary time-constant of the CT, thereby reducing the overdimensioning factor necessary for faithful transformation.

The standard IEC 60044-6 classifies CTs according to transient performance (TP) when short circuit current has exponentially decaying DC offset. This standard differentiates four TP classes depending on the layout of the CT core.

TPS Class

TPS class is a closed iron-core CT with small leakage flux. Behavior is defined by the magnetization curve (knee point voltage, magnetizing current) and the secondary winding resistance. It corresponds to Class X in British Standard BS3938-1973 and PL class in Australian Standard AS1675-1986. It is designed for application in differential protection schemes.

TPX Class

TPX is a closed iron-core CT without limitation of the remanence (no limit for remanent flux).

This construction corresponds to Class P in IEC60044-1 but TPX additionally specifies the TP.

FIGURE 10.9   Linear coupler current transformer.

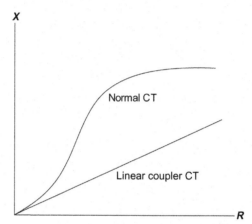

FIGURE 10.10   Linear coupler current transformer (CT) versus normal CT.

TPY Class

These are CTs with antiremanence air gaps (remanence < 10%). Otherwise they behave the same as the TPX class.

TPZ Class

TPZ are CTs with a linear core (remanence may be neglected). The specified transformation accuracy only applies to the AC component while exponentially decaying DC is radically reduced.

In summary:

The IEC classification for TP of CTs as follows:

TPS: Class X in British Standards
TPX: Class P, no remanence limit
TPY: Remanence less than 10% small air gap in core
TPZ: Very low remanence and can be neglected

A linear coupler air gap core with linear characteristics is shown in Fig. 10.9 and is discussed in the following section.

In linear coupler V and $I_{primary}$ has a linear characteristic and are used in voltage balance differential relays used in BB protection.

As an example for comparing the IEC and IEEE CT classes as follows: 5P20 (IEC) → C 100 (IEEE).

10.5.6 Linear Coupler Current Transformers

Linear coupler CTs have an air core and linear characteristics as shown in Fig. 10.10. It produces a voltage that is proportional to the primary current of the CT and can be used in busbar protection voltage differential relays. The main advantage of these CTs is the avoidance of a saturated operating region. When these CTs have a small air gap in an iron core, they are known as called transactors.

10.6 CURRENT TRANSFORMERS CONNECTIONS

Way-connected CTs are shown in Fig. 10.11.
Delta-connected CTs are shown in Fig. 10.12.

10.7 CURRENT TRANSFORMER KNEE POINT

IEC specifies the knee point as the cross section of continuation of the two linear sections of the CT curve (see Fig. 10.13). The ANSI/IEEE method to determine the knee point on the curve is to calculate the point where the tangent to the curve is at 45° degrees to the abscissa (see Fig. 10.14).

10.8 OPTICAL CURRENT AND VOLTAGE TRANSFORMER

A new optical technology was introduced within substations to replace the conventional current and voltage transformers (CT and VT). The advantage of this technology is that it will makes the new devices simple, compact, and reduces the Ferroresonance effect associated with conventional one. It also gives a direct connection digitally to the protection and automation system of substation. The optical current transformer (OCT) and the optical voltage

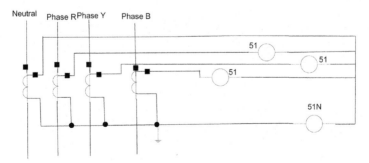

FIGURE 10.11 Way-connected current transformers.

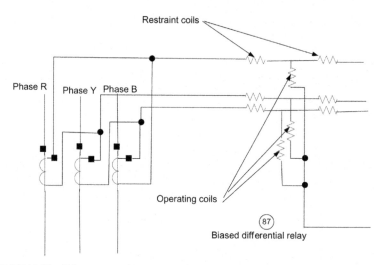

FIGURE 10.12 Delta-connected current transformers.

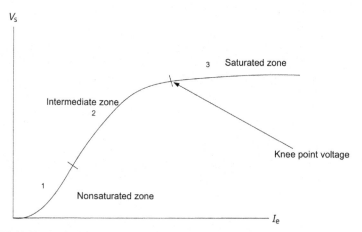

FIGURE 10.13 Excitation curve of current transformer.

transformer (OVT) are used in substations that allow a direct connection to transducers to substation communication network via IEC 61850.

10.8.1 Advantages of Optical Instruments

Optical high-voltage CTs (Fig. 10.15) provide several benefits over conventional CTs and VTs:

Better accuracy: The accuracy of measurment is better in optical instruments.

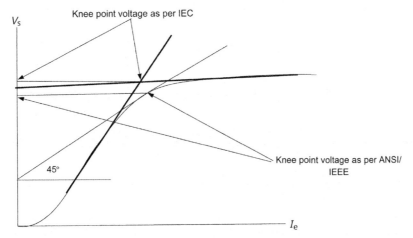

FIGURE 10.14 Methods to determine knee point voltage as per both ANSI/IEEE and IEC methods.

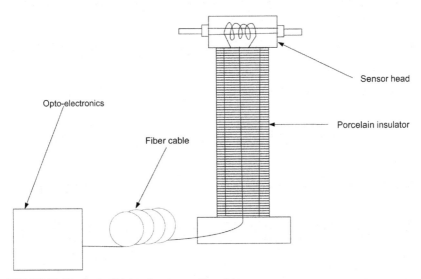

FIGURE 10.15 Optical high-voltage current transformer.

Output bandwidth: The bandwidth of the optical instruments is wider than conventional one.

Combined equipment: One compact unit includes both the CT and the VT.

Environmentally friendly: There is no open circuit hazard for secondary winding of the CT, or Ferro resonance risks.

Applications: As an OCT uses light to measure the magnetic field surrounding a current carrying conductor, an optical measurement of current has the ability to measure fault currents exceeding 400 kA peak. Additionally both AC and DC currents can be measured to this accuracy level.

10.9   CURRENT TRANSFORMER COMMISSIONING TESTING

10.9.1   Visual Checks

The following visual checks should be carried out:

1. Check the ratings on the CT nameplate are as per the approved drawings and specifications.
2. Inspect the CT mechanically for any damage or defects.
3. Check that the CT connections are as per schematic drawings.
4. Check the clearance between the primary circuits and the secondary circuits on the CT especially for if there are any gas-insulated CTs.
5. Check the tightness of all CTs' secondary terminations at the local control cubicle (LCC) panels, at relay panels, and at metering panels.
6. Check that the CT is grounded in a single point at LCC panels or on protection panels as per drawings.
7. Check that the CT has the correct shorting terminal blocks and the ground cable has the correct cross-section area.

10.9.2   Commissioning Electrical Tests

The following electrical tests should be carried out.

10.9.2.1   Insulation Resistance Test

This test is performed to ground of the CT and wiring at 1000 V DC. There should also insulation resistance tests and dielectric-withstand tests on the primary winding with the secondary winding grounded. The voltage will be applied as follows:

- Between primary and secondary connected to the ground.
- Between secondary to primary connected to ground.
- Between secondary cores to cores if the CT has different cores.

With voltage of 1000 V DC applied for 30 seconds first, then up to 60 seconds as mentioned above; investigate any measurements less than 100 M Ohms.

The voltage withstand test is performed at 2000 V AC 50 Hz for 30 seconds between the primary and secondary windings of the CT.

10.9.2.2 Winding Resistance Test

During this test we can measure the secondary resistance of the CT for each core after opening the CT connection to relay panels or metering panels.

10.9.2.3 Polarity Test

This test is performed to confirm the correct marking on the CT primary and secondary as per the drawings. This test can be done by a flick DC test using a DC battery as shown in Fig. 10.16. Close and open the battery switch instantaneously that is connected to the primary CT then check that the pointer is moving in the positive direction of the galvanometer; if the movement is in a positive direction then the polarity is acceptable.

10.9.2.4 Ratio Test by Primary Injection

This test is done to check the CT ratio for all cores as per the nameplate and is done by primary injection as shown in Fig. 10.17.

- Injecting between each phase and neutral to check the ratio between the injected current in the primary side (this should be more than 25% of the rated primary current of the CT) and secondary currents in each core.
- When injecting between each two phases there should be no any current in the neutral.

10.9.2.5 Magnetizing Current Test

This test is done by injecting a voltage on secondary terminals of the CT (Fig. 10.18) and to monitor the excitation current up to the point that the

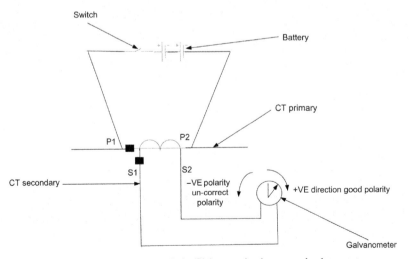

FIGURE 10.16 Current transformer polarity/flick test using battery and galvanometer.

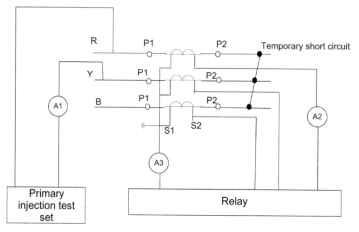

FIGURE 10.17 Primary injection ratio test of a current transformer.

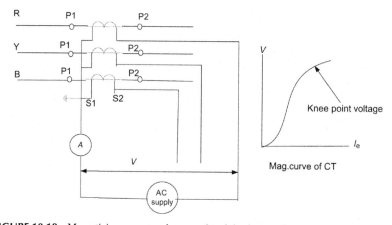

FIGURE 10.18 Magnetizing curve test by secondary injection test for a current transformer.

transformer reaches saturation. This allows you to plot the excitation curve of the CT and compare it with the one provided by the manufacturer.

10.9.2.6 Loop Resistance Burden Test

This test is done to ensure that the connected burden to the CT is within the rated burden, identified on the nameplate.

Inject the rated secondary current of the CT from the transformer terminals towards the load side and isolate the transformer terminals at the CT terminal blocks. Observe the voltage drop across the injection points as the burden can be found from this equation as follows:

VA (burden) = voltage drop × rated CT secondary current

Refer to Fig. 10.19 for the burden test.

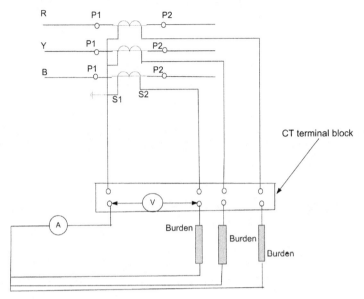

FIGURE 10.19   Current transformer burden test.

10.9.2.7   Continuity of Secondary Circuits

The test should including checking of terminations, shorting links by the primary and secondary injection test.

10.9.2.8   High Voltage Test

This which is undertaken during the high-voltage test at the final stages of commissioning tests, but with disconnecting the CT connections to the relays and shorting the transformer secondaries during the high-voltage test.

10.9.2.9   Demagnetizing the CT Cores

Demagnetizing the CT cores by connecting the transformer's secondaries to the ground at the end of all tests.

Chapter 11

Voltage Transformers

11.1 INTRODUCTION

This instrument is used to transfer the primary voltage to a secondary voltage proportional with the primary voltage. The secondary voltage value is suitable for measuring and protection devices.

Voltage transformers (VTs) are used in:

- Metering;
- Overvoltage protection, for example, no-load line;
- Under voltage protection, for example, overloads line;
- Discharging capacitor banks (wound).

11.2 PRINCIPLE OF OPERATION OF ELECTROMAGNETIC VOLTAGE TRANSFORMERS

The electromagnetic VT is connected across the points at which the voltage is to be measured, and is therefore much like low-power transformers with secondary winding operating close to an open circuit. For such a no-load transformer the voltage transformation is in proportion to the primary (N_p) and the secondary (N_s) turns:

$$\frac{V_p}{V_s} = \frac{N_p}{N_s}$$

An inductive VT is ideally a transformer under no-load conditions where the load current is zero and the voltage drop is only caused by the magnetizing current and is thus negligible. In practice, the winding voltage drops are small, and the rated flux density in the core is designed to be well below the saturation density. Therefore, the exciting current is low and exciting impedance constant with a variation of applied voltage over the operating range including some degree of overvoltage. The parameters that define VT performance are voltage ratio error and phase displacement error. The ratio error is defined as:

$$\text{Ratio Error}\% = \frac{Kn \ Vs - Vp}{Vp} \times 100$$

Practical Power System and Protective Relays Commissioning.
DOI: https://doi.org/10.1016/B978-0-12-816858-5.00011-3
123

where K_n is the rated ratio and V_p and V_s are the primary and secondary terminal voltages. If the error is positive, the secondary voltage exceeds the rated value. The phase error is the phase difference between the secondary and the primary voltage phasors. It is positive when the secondary voltage leads the primary voltage phasor. According to the ratio and angle error, all VTs are required to comply with one of the classes as defined in IEC60044-2, measuring the VT accuracy class (0.1, 0.2, 0.5, 1, and 3) or protective VT accuracy class (3P or 6P).

VTs should be of a sufficient size as to prevent measured disturbances from inducing saturation in the VT. For transients, this generally requires that the knee point of the VT saturation curve be at least 200% of the rated system voltage. It is always good practice to incorporate some allowance in the calculations for overvoltage conditions. The frequency response of standard metering and protection class VTs depend on their type and burden.

In general, the burden should be very high impedance. This is generally not a problem with most monitoring equipment available today. Power quality monitoring instruments, intelligent electronic devices (IEDs), and other instruments all present very high impedance to the VT. With a high impedance burden, the response is usually adequate to at least 5 kHz. While working in energized V.T it is very dangerous to short circuit the secondary winding of the voltage transformer.

Refer to Figs. 11.1A,B,C, 11.2, and 11.3 for examples of VTs.

Refer to Fig. 11.3 for VT equivalent circuit.

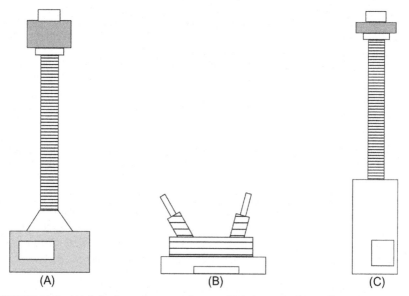

(A)   (B)   (C)

FIGURE 11.1 (A) Inductive voltage transformers (VT) (wound); (B) medium-voltage dual-bushing VT; (C) capacitive VT.

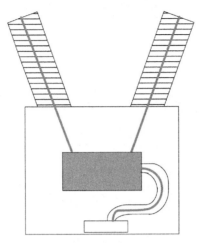

FIGURE 11.2 Cut view of a double-pole voltage transformer.

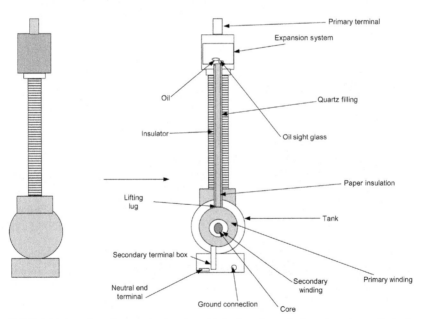

FIGURE 11.3 VT equivalent circuit, where Z_H is the primary leakage impedance; Z_L is the secondary impedance; R_m and X_m are the core losses and exciting components, respectively; and n^2 is the factor to refer Z_H to the secondary side.

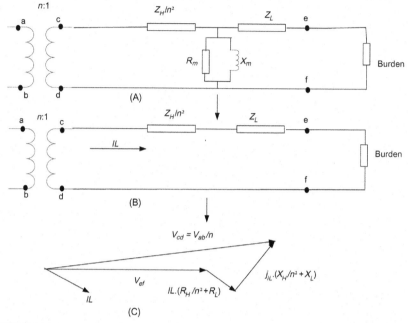

FIGURE 11.4 Inductive VT.

11.3 PRINCIPLE OF OPERATION OF CAPACITIVE VOLTAGE TRANSFORMERS

Due to the cost we use a voltage divide capacitance along with electromagnetic voltage transformer to form in total the capacitive voltage transformer (CVT) as shown in Fig. 11.5.

This reduces the insulation requirements then later; as a result of this cost is also reduced.

The total ratio of $CVT = ((C1 + C2)/C1) \times T$ where T is the electromagnetic voltage transformer ratio. Fig. 11.6 shows the equivalent circuit of voltage transformer.

The capacitance of the voltage divider in CVT and the reactance in electromagnetic VT form a resonance circuit called a Ferroresonance circuit. This circuit when subjected to a step change in voltage leads to an oscillation of nonlinear nature called Ferroresonance which in turn can be dangerous for the VT if it is left for a long time. Earth switch is therefore used to absorb this oscillation in Ferroresonance relay scheme.

11.4 BURDENS AND ACCURACY CLASSES

There are two accuracy classes for protective voltage transformer and metering voltage transformer as per IEC 60044-2 as shown in Table 11.1.

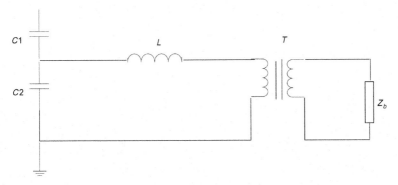

FIGURE 11.5   Capacitive voltage transformer circuit.

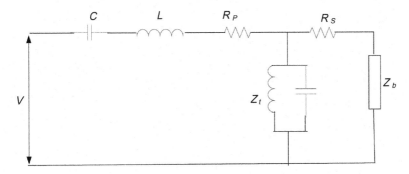

FIGURE 11.6   Capacitive voltage transformer equivalent circuit.

TABLE 11.1 Voltage Transformer Accuracy Classes According to IEC 60044-2

Class	Ratio Error (%)	Phase Displacement Error (min)
0.1	0.1	5
0.2	0.2	10
0.5	0.5	20
1	1	40
3	3	–
3P	3	120
6P	6	240

11.5 TYPES AND THREE-PHASE CONNECTIONS OF VOLTAGE TRANSFORMERS

The voltage transformer connection can be shown in Fig. 11.7.

11.6 OPTICAL CURRENT AND VOLTAGE TRANSFORMERS

A new optical technology was introduced within a substation to replace conventional current and VTs with the advantage that technology being that it makes the new devices simple, compact, and reduces the Ferroresonance effect associated with conventional VTs, and also gives a direct connection digitally to the protection and automation system of the substation. The optical current transformer and the optical voltage transformer are used in substations which allow a direct connection from transducers to the substation communication network via IEC 61850.

11.6.1 Advantages of Optical Instruments

Optical instruments is more suitable and matching to the new digital protection and automation.

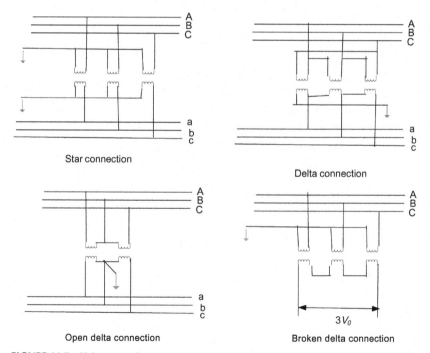

FIGURE 11.7 Voltage transformer connection components.

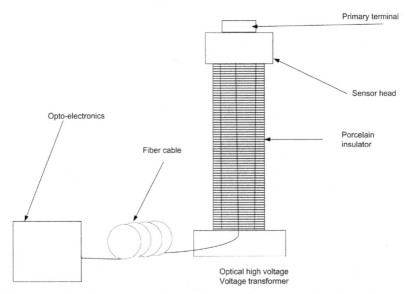

FIGURE 11.8   Optical HV voltage transformer.

Optical sensors (see Fig. 11.8 for an optical high-voltage and current transformer) provide several benefits over conventional CTs, VTs, and CCVTs, including the following:

Better accuracy, with the accuracy being better in optical instruments;
Output bandwidth—the bandwidth of optical instruments is wider than for conventional ones;
Combined CT and VT as a compact unit includes the CT and VT;
Greater safety as they are environmentally friendly, and there is no open circuit hazard for secondary winding of CT or Ferroresonance risks in VTs.

11.7   VOLTAGE TRANSFORMER TESTING

11.7.1   Visual Check

The visual check is carried out as follows:

1. Check that the VT as per the manufacturer-provided drawings and manuals at the site.
2. Check VT nameplate ratings and terminal markings.
3. Check the primary connection is correct as per drawings especially in SF6 gas insulated system (GIS).

4. Check secondary connections in the VT terminal box and in the local control cubicle (LCC), the tightness and cross-sectional area of the cables, and the color codes for phases and the ground cable.
5. Check the oil level in the outdoor PT and SF6 pressure in the SF6 GIS.
6. Check the MCB ratings of the VT secondary terminals in the LCC.

11.7.2   Insulation Resistance Test

This test is carried out at 5000 V DC as follows:

1. between primary and secondary;
2. between primary and earth; and
3. between secondary and earth.

Also, the voltage withstand test is performed at 2000 V AC 50 or 60 Hz for 30 seconds between the primary and secondary windings of the VT.

11.7.3   Polarity (Flick) Test

This test is performed to confirm the correct marking on the VT primary and secondary as per the drawings, and it can be done by a flick DC test using a DC battery as shown in Fig. 11.9. The battery switch which is connected to the PT primary is instantaneously closed and opened and then the pointer is

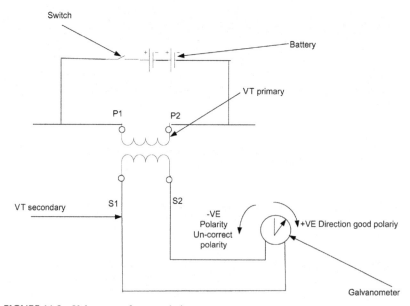

FIGURE 11.9   Voltage transformer polarity test.

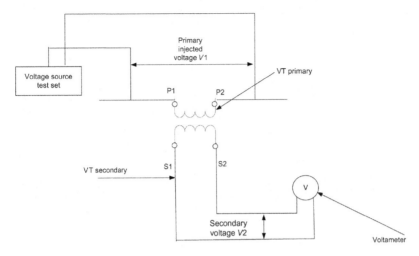

Voltage transformer ratio = $V1/V2$

FIGURE 11.10 Voltage transformer ratio test.

checked for direction of movement in the galvanometer and if it is in the positive direction then the polarity is ok.

11.7.4 Voltage Transformer Ratio Test

A variable AC voltage source advanced test set is used to inject voltages on the primary side and the secondary voltage is measured to check that the VT ratio is correct for each phase, as shown in Fig. 11.10.

11.7.5 Winding Resistance Test

Using this test we can measure the secondary resistance of the VT for each core at the VT terminal box and at the LCC panels between phases and the neutral, and between phases and other phases, after opening the VT connection to relay panels or metering panels by opening all the VT secondary MCBs.

11.7.6 Loop Resistance Burden Test

This test is done to ensure that the connected burden to VT is within the rated burden, identified on the nameplate. Using a secondary injection tester to inject the rated secondary voltage of the VT from the VT terminals toward the load side while isolating the VT terminals at the VT terminal blocks toward the VT and observe the voltage drop across the injection points as the burden can be found from the following equation:

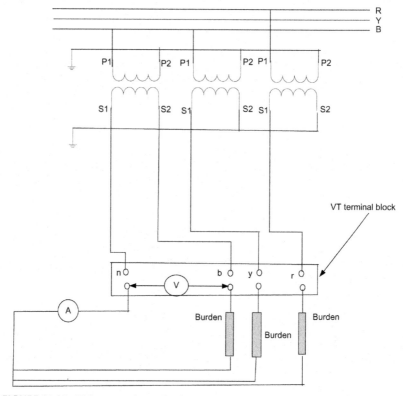

FIGURE 11.11   Voltage transformer burden test.

VA(burden) = voltage drop × rated VT measured secondary current

Refer to Fig. 11.11 for the VT burden test.

Chapter 12

Disconnecting Switches and Earthing Switches Theory Testing and Commissioning

12.1 INTRODUCTION

A disconnect switch (DS, also known as a disconnector or isolator switch) is a mechanical device used to isolate the equipment undergoing maintenance, such as a circuit breaker, in a safe manner. It can also be used to transfer the circuit bay from one busbar to another busbar, for example, in a bus coupler circuit in a substation.

These devices can be provided with a safety earthing switch (ES) for the safety of the personnel who will work in the electric circuit within a substation or power station yards.

Whilst a DS can carry load current, it can only disconnect or close the circuit in a no-load condition; there is an interlocking circuit in substations to prevent them from operation under load conditions.

Mechanisms are normally installed to permit the operation of the DS by an operator standing at ground level. The operating mechanisms provide a swing arm or gearing to permit operation with reasonable effort by utility personnel. Motor operating mechanisms are also available and are applied when remote switching is necessary.

The DS operation can be designed for vertical or horizontal operating of the switch blades.

Several configurations are frequently used for switch applications, as seen in Figs. 12.1–12.7. These configurations include the following:

- Horizontally-upright mounted DS
- Vertically-mounted DS
- Double break switches
- V switches
- Hook-stick switches
- Pantograph switches
- Earthing switches

Practical Power System and Protective Relays Commissioning.
DOI: https://doi.org/10.1016/B978-0-12-816858-5.00012-5

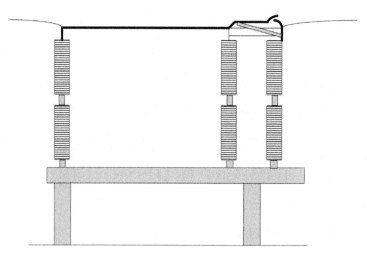

FIGURE 12.1 Horizontally-upright mounted disconnect switch.

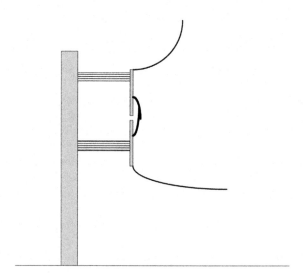

FIGURE 12.2 Vertically-mounted disconnect switch.

12.1.1 Load Break Switches

A load break switch is a type of DS that can open or close on specified load currents. It is normally used to energize or de-energize a circuit that has some magnetic or capacitive currents as a transformer excitation current and high-voltage transmission cables with capacitive current.

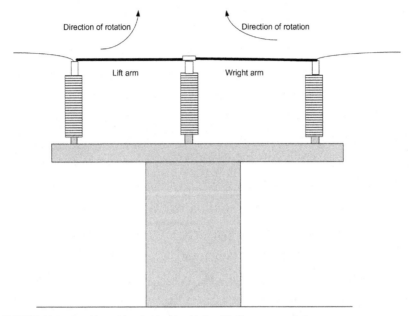

FIGURE 12.3 Double end break (double side break) disconnect switch.

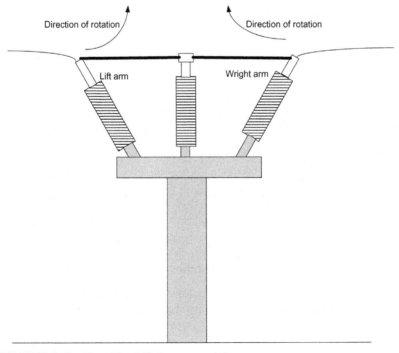

FIGURE 12.4 Double end break V-disconnect switch.

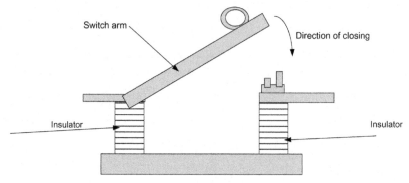

FIGURE 12.5 Hook-stick operated disconnect switch.

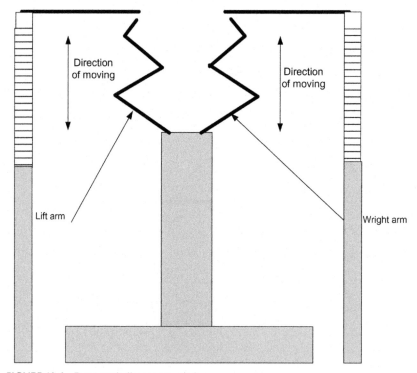

FIGURE 12.6 Pantograph disconnect switch.

12.1.2 High-Speed Earthing Switches

High-speed ES are used to discharge the high voltage cables from the capacitive charging current. This current is very dangerous for the personnel working at the high-voltage cable end at a substation, paticularly if they are outdoors.

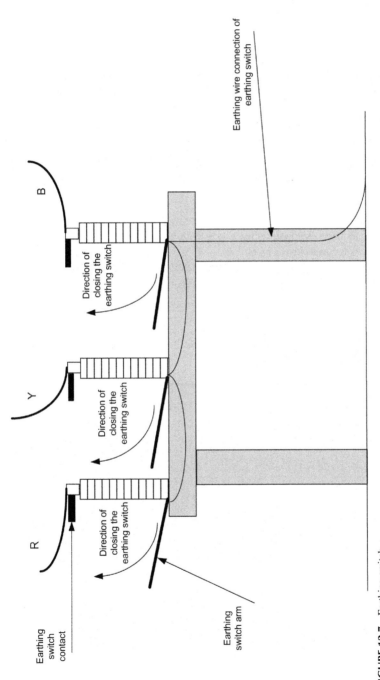

FIGURE 12.7 Earthing switch.

12.2 DISCONNECT SWITCHES/EARTH (GROUNDING) SWITCHES TESTS

This includes the following tests:

1. Mechanical check and function test including open/close and electrical and mechanical interlocks

   a. Check the horizontal level of the disconnecting switch in the open and close positions.

   b. Measure the distance between the axis of the two isolators of the disconnecting switch as per the drawings.

   c. Check that the mechanical indicator of the disconnecting switch is correct for the open and close positions.

   d. Check for the mechanical interlocks between the disconnecting switch and its ES.

2. Operating time test

   a. This test can be done using a stop watch.

3. Contact resistance test

   a. This test is done by micro-Ohm meter tester which injects 100 DC Amp and is connected as shown in Fig. 12.8.

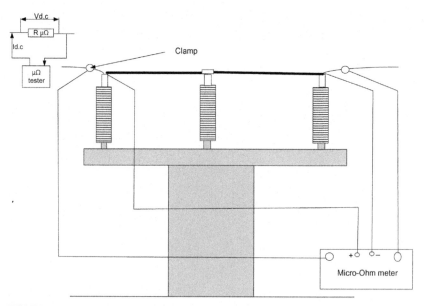

FIGURE 12.8 Contact resistance measurement of the disconnecting switch.

The results at site shown in Fig. 12.8 was in the range of 45−50 μΩ.

4. Insulation test
   a. This can be done by Megger tester at 5000 V DC.
5. High voltage AC test
   a. This test will be done during the first energization of the substation with a nominal voltage of the disconnecting switch.

Chapter 13

Fault Recorders in Substations and Power Stations

13.1 INTRODUCTION

In the past there was a need to have a separate fault recorder in substations and at power stations. In some cases there is now a recorder included in the new numerical protection relays. However, there is still a need for a separate fault recorder when we need a detailed information for prefault, fault, and postfault conditions. Retaining this information requires a large computer memory, which is only found in separate units.

The inputs for these fault recorder units comes from the voltage transformer and the current transformer. The data shows the analog signals Vr, Vy, Vb, Ir, Iy, and Ib. It also shows the operation of the protection relays and tripping relays.

Another device available is the sequence of event recorder which shows the time of operation of substation components such as as circuit breakers and disconnect switches.

All these devices are very important in analyzing faults within a power system and to predict any faults that may occur with the system.

For any event in the fault recorder there are three important time sequences:

- Prefault time duration
- Fault time duration
- Postfault time duration

A fault recorder unit can be used to adjust these times, should they occur, and there is also a setting for which digital channels trigger the recording process.

A fault recorder consists of the following units:

- Data acquisition units that have inputs from the current transformer, voltage transformer, analog channels, and digital channels.
- Restitution unit which collect the information from data acquisition units.
- Local computer unit.
- Local printer.

Fig. 13.1 shows a setup for a fault recorder unit.

Practical Power System and Protective Relays Commissioning.
DOI: https://doi.org/10.1016/B978-0-12-816858-5.00013-7

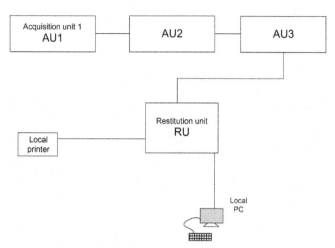

FIGURE 13.1   Fault recorder unit setup.

13.2   FAULT RECORDER TESTING

These tests include the following:

1. Adjusting the prefault, fault, and postfault times on the fault recorder units and check that they are working correctly for a simulated fault by secondary injection. Check the operation of both the analog channels and the digital channels.
2. Check that all points of analog channels and digital channels are transferred to the local and remote annunciator inside the substation.
3. Check that all points of analog channels and digital channels are transferred to the National Control Center SCADA system.

Chapter 14

Gas Insulated System Substations

14.1 INTRODUCTION

The gas insulated system (GIS) has been used since 1960, and consists of tubes that contain SF_6 under a pressure of about 6 bars. This gas, under pressure, provides very good insulation for high voltage and extra-high voltage systems. Fig. 14.1 shows the GIS system components.

This system can be either a single three-phase enclosure, for example in a high voltage (66 kV) system or a separate single phase enclosure, for example in extra-high voltage (400 kV and 500 kV) systems, seen in Fig. 14.2. The latter system is more expensive but safer and better for maintenance.

A GIS consists of the following components in each bay or diameter. Each will be discussed below:

- Circuit breakers (CB).
- Current transformers (CT) and voltage transformers (VT).
- Disconnecting switch.
- Earthing switch and high-speed earthing switch.
- Busbars: feeder (bay) or diameter.
- Cable sealing end or GIS bushing for transmission lines.

14.2 DISCONNECTING AND EARTHING SWITCHES

Both the disconnector switch and earthing switch are driven by a DC motor. However, the earthing switch can be operated manually in the event of an emergency.

The high-speed earthing switch is used to earth the system to discharge the line from capacitive charge.

In most installations the disconnections and the earthing switch are in one unit and have mechanical interlocking with each other, as seen in Fig. 14.3.

Practical Power System and Protective Relays Commissioning.
DOI: https://doi.org/10.1016/B978-0-12-816858-5.00014-9

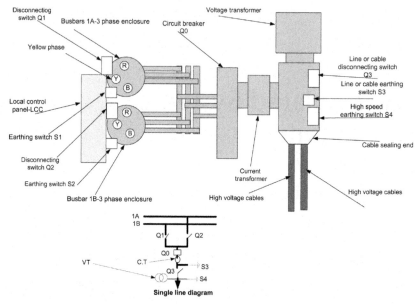

FIGURE 14.1 SF6 gas insulated system components.

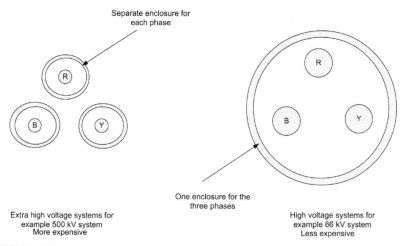

FIGURE 14.2 Three-phase enclosure gas insulated system (GIS) and single phase enclosure GIS.

14.3 CIRCUIT BREAKERS

Similar to air insulated system CB, this device can disconnect or close the circuit under normal and short-circuit conditions. In an SF_6-type CB the quenching medium of the electric arc during circuit opening is the SF_6 under pressure. There are two types of SF_6 CB: a puffer type that is used for high

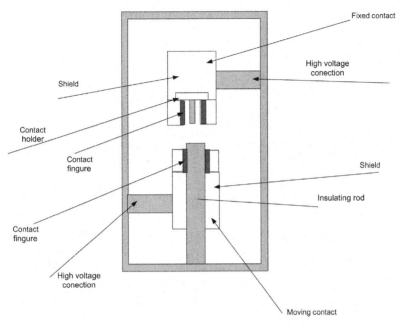

FIGURE 14.3 Gas insulated system disconnect switch.

breaking capacity, and a self-blast type that is used for medium breaking capacity.

14.3.1   Puffer Circuit Breakers

In this type of CB, there are two parts: a moving contact and a fixed contact. Both have a nozzle shape. Also involved is a blast cylinder. During the opening process, the moving contact moves and the volume of the blast cylinder is reduced. This increase the SF 6 gas pressure in the blast cylinder,which is released in the Wright moment to blow the arc and extinguish it as shown in Fig. 14.4.

This type of CB is used in 400 and 500 kV GIS systems.

14.3.2   Self-Blast Circuit Breakers

In this type of CB, the SF_6 gas in the breaking chamber is injected by means of a piston mechanically coupled to the mobile contact. The piston compresses the gas at a high pressure to blow the arc at the zero crossing of the current waveform. See Fig. 14.5.

Compared to the puffer type, the energy required in a self-blast to blow the arc is reduced by 80%, but it is used only in a medium breaking capacity in 132 kV GIS systems.

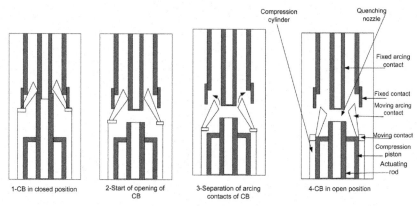

FIGURE 14.4 Puffer type circuit breaker.

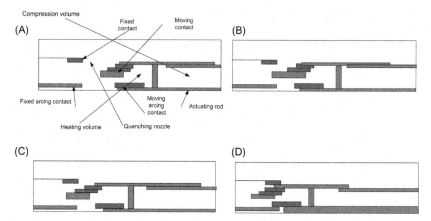

FIGURE 14.5 Self-blast circuit breaker (CB). (A) CB in open position. (B) Interruption of small currents. (C) Interruption of short-circuit currents. (D) CB in closed position.

14.4 INSTRUMENT TRANSFORMERS

These devices transform the current and voltages to standardized levels for protection and control devices to 1 or 5 A for CT and to 100, 110, or 120 V for VT.

In modern substation and bay control systems, current and VT can be replaced by sensors. They offer the same accuracy as conventional instrument transformers. The output signal, converted from analog to digital, is processed by the digital bay control unit, using fiber optics. This is discussed further in Chapter 10, Current Transformers and Chapter 11, Voltage Transformers, of this book.

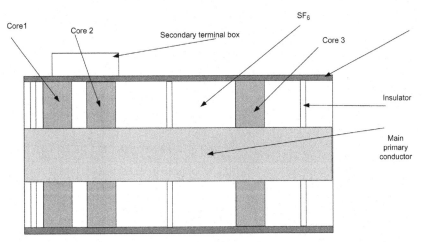

FIGURE 14.6   Gas insulated system current transformer.

14.4.1   Current Transformers

CT are inductive ring types installed either inside the GIS enclosure or outside the GIS enclosure Fig. 14.6 shows the CT for a GIS.

14.4.2   Voltage Transformers

VT are one-phase (phase-to-ground connection) and can be inductive or capacitive. VT can be removed so that the GIS can be high-voltage tested without damaging the transformer. Alternatively, the VT may have a disconnect switch or removable link.

Fig. 14.7 shows an inductive VT for a GIS and Fig. 14.8 shows the capacitive VT details for a GIS.

14.5   CABLE CONNECTION

A cable connecting to a GIS is provided with a cable termination kit. This is installed on the cable to provide a physical barrier between the cable dielectric and the SF_6 gas in the GIS (see Fig. 14.9).

14.6   DIRECT TRANSFORMER CONNECTIONS

To connect a GIS directly to a transformer, a special SF_6 to oil bushing that mounts on the transformer is used as shown in Fig. 14.10. The bushing is connected under oil on one end to the transformer's high-voltage leads. The other end is connected to the SF_6.

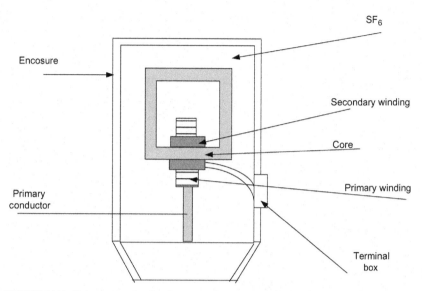

FIGURE 14.7   Inductive gas insulated system voltage transformer.

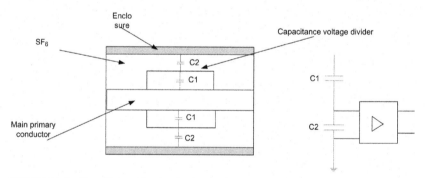

FIGURE 14.8   Capacitive gas insulated system voltage transformer.

14.7   SURGE ARRESTER

Zinc oxide surge arrester elements suitable for immersion in SF_6 are supported by an insulating cylinder inside a GIS enclosure section. This makes a surge arrester for overvoltage control as shown in Fig. 14.11. Insulation coordination studies usually show there is no need for surge arresters in a GIS, however some users specify a surge arrester at transformer and cable connections. Predominantly, the outdoor bushing surge arresters are used.

14.8   CONTROL SYSTEM

The local control of the GIS components (CB, disconnectors, and earthing switches) is undertaken through a local control cabinet (LCC). Additionally, all interlocking and annunciator and alarms are controlled via the LCC.

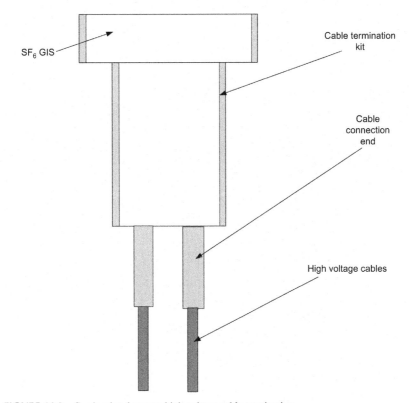

FIGURE 14.9 Gas insulated system high voltage cable termination.

In the 1990s the LCC was built in a separate panel in front of the GIS, an arrangement which can be viewed as being more safe and efficient. However, in more recent installations of SF_6 GISs, the LCC was built above the GIS as shown in Fig. 14.12.

14.9 GAS MONITORING SYSTEM

The insulation and the ability of the SF_6 to isolate and blow the arc in CB is dependent upon the density of the SF_6 gas which is measured by a mechanical temperature-compensated pressure switch to monitor the SF_6 in the gas compartments. The first alarm (Stage 1) indicates that a refill of SF_6 gas in the compartment which has the leakage is required. The second stage trips the CB to block its operation until repair maintenance is undertaken in this compartment. Because it is much easier to measure pressure than density, gas monitoring systems usually have a pressure gauge. A chart is provided with the system to convert pressure and temperature measurements into density.

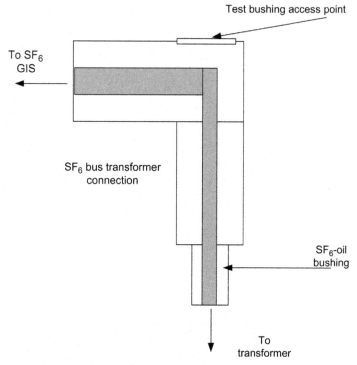

FIGURE 14.10   Transformer to gas insulated system connection.

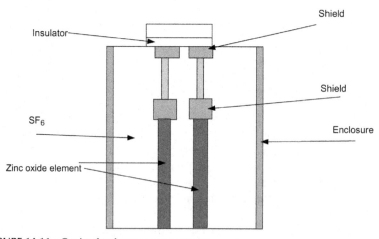

FIGURE 14.11   Gas insulated system surge arrester.

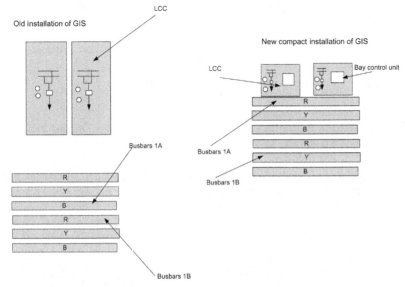

FIGURE 14.12 Local control cabinet (LCC) for gas insulated system.

14.10 GAS COMPARTMENTS AND ZONES

A GIS is divided by gas barrier insulators with a conical shape into gas compartments for gas handling purposes. In some cases, the use of a higher gas pressure in the CB than is needed for the other devices, requires that the CB is a separate gas compartment. Consequently, the GIS is divided into relatively small gas compartments.

The SF_6 gas alarms for each compartment are grouped to one alarm in control room. After an alarm is raised, or for maintenance requirements, personnel must inspect the GIS to locate which compartment has the leakage. Another important issue to keep in mind is that some manufacturers use a plug-type electrical connection from the Current Transformer and Potential Transformer in the GIS to the LCC which is not safe and not recommended.

14.11 GIS TESTING

All GIS equipment should have been tested at the factory as per IEC 517.

The purpose of the tests at site prior to commissioning is:

1. To detect any rare possible damage or anomaly that may occur during transport.
2. To test the complete GIS installation after final assembly at site.
3. To ensure uninterrupted operation in service.

The tests of GIS are done as follows:

- Inspection after transport
- Visual check the GIS for completeness, obvious transport damage, etc. well before commencement of the installation.

The following tests should be undertaken:

1. Measurement of voltage drop

    The voltage drop test is done by injecting 100 Amp DC and measuring any voltage drop in different locations of the system.

2. Check of gas tightness

    All flanges and any connecting parts of the GIS compartments should be checked for their tightness.

3. Contact resistance test for main circuit

    This test is performed at different points in the GIS and the results are compared to the one assembled and measured at factory.

4. SF_6 leakage test

    In this test we put a plastic cover around each compartment of the GIS and leave it for 24 hours then check for any leakage using a SF_6 gas detector. If there is any leakage then investigate the source of this leakage and rectify it before putting the system in service.

5. Insulation test of control circuits

    All control cables and connections in GIS should be checked using a Megger at 2000 V DC for 1 minute and the insulation should be more than 100 mega-Ohms.

6. Dew point measurement

    The dew point measurement is performed once the gas compartments have been filled and pressurized with SF_6 to its correct pressure. In order to allow an equalization, this measurement should not be carried out immediately after the filling. This test should be undertaken for each SF_6 compartment.

7. SF_6 pressure gauge meter test

    In this test we will activate the operation of the alarm contact and trip contact of the SF_6 gauges in each compartment and check that the signals are transferred to LCC and/or central control room if it exists.

8. Time testing

    In this test we measure the opening and closing time of all CB and compare the results with those supplied by the manufacturer at both nominal DC voltage and at reduced DC voltage. This is done for each phase and to measure any difference in opening and closing time for the CB phases that is not detected by phase discrepancy protection off the CB. The closing and opening time of disconnectors and earthing switches are measured and compared with the specifications supplied by manufacturer.

9. Instrument transformer tests
   - Polarity test: this test is done to check the correct polarity of the CT and VT.
   - Ratio test for CT and VT.
   - Voltage withstand test at 2 kV for 30 seconds for CT and VT.
   - Winding resistance test for CT and VT.
   - Loop resistance or burden test for CT and VT.
   - Magnetizing curve test for CT.
   - Insulation resistance test by Megger for CT and VT—both primary and secondary.

10. High voltage test for GIS

This test is done by using cascaded resonance instruments at a specified voltage:

Vtest = 0.8V of the rated voltage, with frequency about 45−300 Hz.

With all CB and disconnectors closed and all earthing switches opened, all CT's secondary Winding are to be short circuited and all VT miniature CB to be opened. A Megger test should also be undertaken after this test.

14.12 SF$_6$ GAS HANDLING

14.12.1 SF$_6$ Gas Filling Cylinder

Fig. 14.13 shows how to fill the GIS compartments when the gas density is less than the desired one for that compartment due to SF$_6$ leakage.

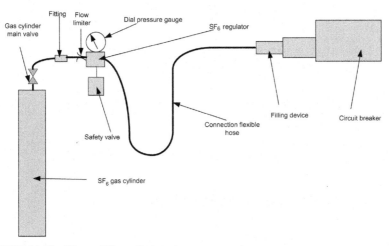

FIGURE 14.13 SF$_6$ gas filling cylinder setup.

14.12.2 Gas Service Truck

All SF_6 equipment must be filled initially and after a time may need refilling. To perform overhaul or maintenance on this equipment the gas must be removed. Releasing the gas into atmosphere would be very expensive and possibly dangerous if done indoors. Most SF_6 gas service trucks are equipped to provide storage, purification, and transfer of gas to and from equipment (Fig. 14.14). The primary functions of an SF_6 gas service truck are to remove the SF_6 from a switchgear or CB, store it, clean and dry it, and return it to the switchgear or CB. The major components required to achieve this are:

- Vacuum pump, for pulling SF_6 out of the switchgear, and for creating a vacuum prior to refilling the switchgear with SF_6.
- An SF_6 storage tank.
- A compressor for pumping SF_6 into the storage tank or back into the switchgear.
- A filtering system for removing moisture and other contaminants from the gas prior to its return to the switchgear.

These major components are interconnected by a series of valves and pipe work in such a way that the gas truck can be operated in several different modes. Several important functions of the truck are monitored. Before filling the equipment, the ambient temperature must be measured and the pressure, temperature, and density chart consulted to determine the correct operating pressure for the temperature.

Every effort must be made to prevent air from entering and mixing with the gas, by means of purging or evacuating portable hoses, and by correct valve selection.

When equipment has been open to the atmosphere, it must be evacuated to remove air and any absorbed moisture before refilling. After filling the

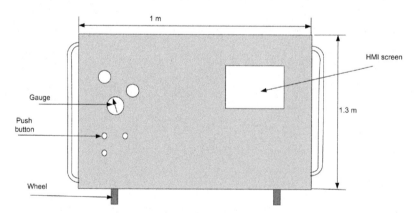

FIGURE 14.14 SF_6 gas filling truck. *HMI*, human machine interface.

equipment it should be left to stabilize for several hours to ensure the density is correct.

The truck system consists of the following components:

- Oil-free suction pump for recovery of SF_6.
- Vacuum pump for evacuation of air.
- Operating indication of the most important process parameters in a screen panel.
- Evaporator.
- Particle filter.
- Dry filter.
- Pressure and vacuum units.
- Weighing scales for SF_6 bottles.
- Long storage connecting hose.

Note: Before any gas handling is performed it is important to ensure that the equipment has been properly isolated from energized electrical sources.

Chapter 15

Batteries

15.1 INTRODUCTION

15.1.1 Power Sources

AC power is often preferred because it is easy to generate, to transmit, and to use. However, in some situations, DC power is preferred because it can be stored in batteries and used if the generating equipment fails.

15.1.2 DC Power Circuits

A DC power circuit consists of a battery charger with an AC supply, a battery, and loads. The battery charger converts AC to DC. The resulting DC is used to supply the load and to supply a charge to the battery (float charge). The battery stores DC power and is used to supply power to the load if the AC supply and charger fail; if the demand is high, the battery may also supplement the power supplied by the charger.

15.1.3 Cells and Batteries

Cells are devices that store and deliver DC power. A battery is made up of one or more cells connected in series. Each cell of an industrial storage battery provides a specific voltage; the battery voltage depends on the number of cells. The cells are filled with an electrolyte (a solution of acid and water) that activates a chemical reaction and provides a path for current flow. The plates are connected to the appropriate terminal posts (positive or negative, depending on the plate material). Many plates may be connected together to increase the storage capacity of the cell. An outer jar contains the plates and the electrolyte.

15.1.4 Cell Operation

Connecting a battery charger to the terminals of a battery starts a chemical reaction, causing electrons to leave one plate and build up on the other. One plate becomes positively charged, and the other is negatively charged. If a load is connected across the terminals, current flows and the stored charge is

Practical Power System and Protective Relays Commissioning.
DOI: https://doi.org/10.1016/B978-0-12-816858-5.00015-0

dissipated. The charge slowly dissipates from charged cells over a period of time. Keeping the cells continuously connected to the charger ensures they will be fully charged at all times.

15.1.5   Safety Considerations

The charge stored in the cells can cause shocks, even if the battery has been disconnected. Electrolytes can cause burns. The hydrogen given off by the cells can cause explosions. Battery rooms should be kept clean, well-ventilated, and locked. When working on batteries, electricians should wear protective gloves, aprons, boots, goggles, and face shields.

15.1.6   Additional Checks

Safety equipment, such as eyewash stations or showers, should be located and checked to ensure proper operation. The temperature of the room should be verified. The battery racks and the floor under the racks should be checked for signs of acid spills. Spills should be reported and cleaned, following facility procedures.

15.1.7   Acid Concentration

The initial concentration of acid in a cell is determined by the manufacturer. When a cell is discharged, acid concentration is low; when a cell is charged, acid concentration is high. The concentration of acid in the electrolyte is a good indication of the amount of charge in the cell.

15.1.8   Specific Gravity and Acid Concentration

The specific gravity of a liquid indicates how heavy the liquid is when compared with pure water. The specific gravity of electrolyte depends on the concentration of acid. Therefore, the specific gravity can be used to determine the amount of charge in the cell. Specific gravity measurements must be corrected for temperature and level of the electrolyte.

15.1.9   Determining the Condition of a Battery

The final determination is usually made by facility engineers or supervisors based on data usually supplied by the electrician. Two measurements are needed: the corrected specific gravity measurement and the direct voltage measurement.

15.1.10   Taking Measurements

A voltmeter may be used to check cell voltage. A hydrometer may be used to measure specific gravity. In general the batteries are used in power stations and substations to supply a DC current for tripping circuits and protection relays and can be found in 110 or 220 V and for communication equipment of 24 or 48 V. An alkaline battery consists of cell container, positive plates, negative plates, and ebonite separators. Lead-acid batteries also exist.

15.2   CHARGING AND DISCHARGING OF A NEW BATTERY

To start a new battery, the following steps are required:

Clean the surface of the batteries; check all positive and negative connections polarities.

Check all tightness of all joints between the battery cells.

Open the battery filling caps then fill the battery with the electrolyte with specific gravity between 1.150 and 1.200.

Leave the battery about 10 hours for plates to be saturated with electrolyte.

Measure the electrolyte level which should be above the plates by 12−15 cm.

Connect the battery to battery charger.

Start the charging with the charger in fast-charging mode with a current equal to battery Amp hour (Ah), for example, for 400 Ah the charging current will be 40 A for 10 hours.

On the beginning of the charge start the charge with 10 A for 30 minutes, and then increase the current to 40 A in steps.

When battery reaches 2.35 V/cell, decrease the charging current to 30 A.

After 60 hours decrease the charging current to 20 A.

When battery reaches 2.6 V/cell stop charging, then restart charging at 10 A until the voltage of cells are stabilized and no changes occur for 2 hours.

Put the battery charger in floating mode of charge.

Charging or discharging should not be undertaken if the temperature gets to greater than 45°C−55°C. For example, for a 400 Ah 220 V lead battery the following measurements were taken:

Lowest cell = 1.73 V/cell

Number of cells in the battery = 164 cells of lead-acid

Highest cell = 2.6 V/cell

Highest temperature: 55°C—hence stop charge or discharge immediately; let the battery temperature decrease over time.

Complete the charging cycle of the battery: first charge, discharge, second charge.

Start the discharging process

15.2.1 Discharging of a New Battery

- Leave the battery to discharge in an external resistance (water resistance may be used at site) for about 2 hours.
- For example: The discharge current should start at 66 A for a 400 Ah 220 V battery for 200 ms, then 49 A for 2 hours.

15.3 BATTERY CHARGER

A battery charger consists of a rectifier circuit, power circuit, ripple monitoring, control circuit, regulator circuit, and fault detection circuit. This charger can also be used as a DC source for a control and protection circuit of a substation during normal operation, or to charge the battery in floating mode. When there is a problem in the AC system, then the battery supplies the DC loads in a substation. There are two types of charging modes: the first is the fast charging for a new or unused batteries, and the second is the floating charge to charge the batteries in service and supply a load to compensate for the small charge lost by the battery in service. Fig. 15.1 shows a DC system using batteries in a high-voltage substations.

15.4 CHARGER SETTING MODES DURING BATTERY CHARGING

Depending upon the cell voltage, the charger can be set in the following modes as seen in Fig. 15.2.

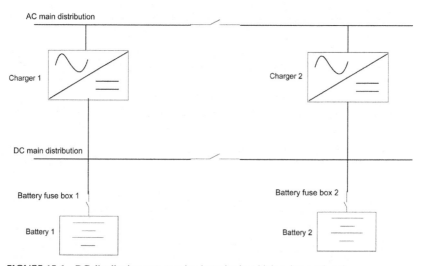

FIGURE 15.1 DC distribution system using batteries in a high-voltage substation.

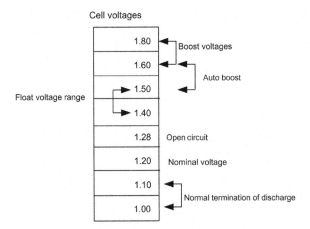

FIGURE 15.2   Operating voltage range for battery charger settings for the charging of alkaline batteries.

The boost charging means the commissioning charging mode at the beginning of new battery operation. The floating charge is the charging of the battery at normal service.

15.5   BATTERIES IN HIGH-VOLTAGE SUBSTATIONS

For a 400/220/132/33 kV substation, the specifications are:

- The minimum ampere-hour ratings of battery for 400 kV switchyard of 400/220/132/33 kV substation and all 400 kV equipment shall be 300 Ah.
- The voltage rating of an 300 Ah battery shall be 220 ± 10% V.
- For 220, 132, and 33 kV switchyard portion of 400/220/132/33 kV substation and equipment, the minimum Amp-hour rating of battery shall be 400 Ah.
- The voltage rating of a 400 Ah battery shall be 220 ± 10% V.
- The DC system shall consist of two battery chargers of 220 V, 300 Ah for 400 kV and battery charger of 220 V, 400 Ah for 220/132/33 kV.
- For 400 kV there shall be two sets of battery with an amp-hour rating of 300 Ah and for all other voltage classes, one battery set with a rating of 400 Ah.

For a 220/132/33 kV substation, the specifications are:

- The minimum ampere-hour rating of battery shall be 300 Ah.
- The voltage rating of a 300 Ah battery shall be 220 ± 10% V.
- Two battery chargers of 300 Ah shall be required.

For a 132/33 kV substation, the specifications are:

- The minimum ampere-hour rating of battery shall be 200 Ah.
- The voltage rating of 200 Ah battery shall be 220 ± 10% V.
- There shall be one battery charger of 200 Ah.

The trickle charge and quick charge voltage per cell of the above batteries shall be $2.15-2.25 \pm 0.02$ V and $1.85-2.75$ V respectively. The range of charging current of the batteries shall be as follows:

- For 300 Ah battery: Boost charge 42−21 A and float charge 720−240 mA
- For 400 Ah battery: Boost charge 56−28 A and float charge 960−320 mA
- For 200 Ah battery: Boost charge 28−14 A and float charge 480−160 mA

The minimum demand load on the chargers are:

- Charger for 300 Ah battery: 24 A
- Charger for 400 Ah battery: 40 A
- Charger for 200 Ah battery: 16 A

Chapter 16

Power System Fault Analysis

16.1 BASIC SYSTEM RELATIONSHIPS

S = total power = $\sqrt{3}\, V_L\, I_L$
P = real power = $\sqrt{3}\, V_L\, I_L \cos(\Phi)$
Q = reactive power = $\sqrt{3}\, V_L\, I_L \sin(\Phi)$
V_L = line voltage, $V_L = \sqrt{3}\, V_{PH}$, V_{PH} = phase voltage, $\cos(\Phi)$ = power factor as shown in Fig. 16.1 which describes the relation between the power system quantities in vector diagram representaion.

Operator a:

$$a = 1\angle 120°,\ a^2 = 1\angle 240°,\ 1 + a + a^2 = 0$$

Operator J:

$$j = 1\angle 90°,\ j^2 = -1$$

$MVA_{base} = \sqrt{3}\, kV_{base} \times I_{base}$. The common usage is 100 MVA as base MVA

$$Z\% = Z(\Omega) \times \frac{MVA_b}{(Kv)^2}$$

$$Z\% = \frac{MVA_b}{MVA_f} \times 100,\ \text{where } MVA_F = \text{fault MVA}$$

16.2 SYMMETRICAL AND UNSYMMETRICAL COMPONENTS

Any unsymmetrical phasors can be analyzed as three symmetrical components, as shown in Fig. 16.2.

16.2.1 Positive Sequence

The three phasors are equal in magnitude. They are displaced from each other by 120 degrees in phase, and have the same phase sequence as the original phasors.

Practical Power System and Protective Relays Commissioning.
DOI: https://doi.org/10.1016/B978-0-12-816858-5.00016-2

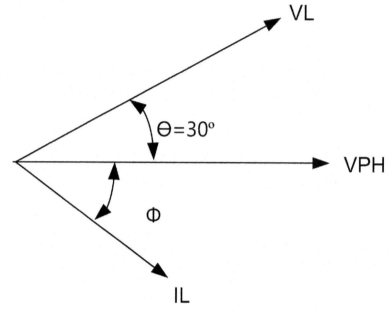

FIGURE 16.1 Voltage and current phasor diagram.

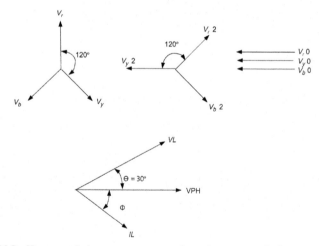

FIGURE 16.2 Unsymmetrical component converted to a three symmetrical components.

16.2.2 Negative Sequence

The three phasors are equal in magnitude. They are displaced from each other by 120 degrees in phase, and have a phase sequence opposite to that of the original phasors.

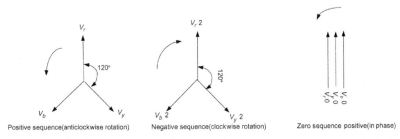

FIGURE 16.3 Positive-, negative-, and zero sequence representation of three-phase vectors.

16.2.3 Zero Sequence

The three phasors are equal in magnitude and with zero phase displacement from each other as shown in Fig. 16.2.

Based on this analysis any faulted network can be represented by three phase networks:

1. Positive-phase sequence network;
2. Negative-phase sequence network;
3. Zero-phase sequence network.

The advantage of this representation is that the three-phase network can be represented in a single-phase network and analyzed simply for faults, as shown in Fig. 16.3.

16.2.4 The Symmetrical Components of Unsymmetrical Phasors

A set of three unbalanced phasors (V_a, V_b, and V_c) can be represented as three sets of phasors: a positive-sequence set, a negative-sequence set, and a zero-sequence set, where:

$$V_{a1} = \frac{1}{3}(V_a + aV_b + a^2V_c)$$

$$V_{a2} = \frac{1}{3}(V_a + a^2V_b + aV_c)$$

$$V_{a0} = \frac{1}{3}(V_a + V_b + V_c)$$

16.3 SEQUENCE IMPEDANCE NETWORKS

Refer to Fig. 16.4 for a sequence network representation.

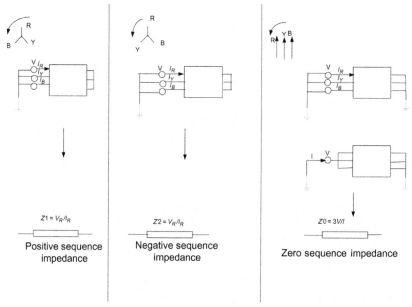

FIGURE 16.4 Sequence network representation.

16.3.1 Power Transformer, Generator, Cables, Transmission Lines Sequence Networks

A power transformer can be represented as $Z\%$ between a high- and low-voltage terminal regardless of voltage level, as shown in Fig. 16.5.

Refer also to Figs. 16.6 and 16.7, respectively, for two-winding and three-winding transformer sequence network representations.

Refer to Fig. 16.8 for an autotransformer with tertiary winding sequence networks.

Refer to Fig. 16.9 for earthing transformer sequence networks:

16.3.1.1 Conclusion

Transformer sequence impedances are described here.

Positive-sequence impedance:

$Z_1 = Z\%$ (transformer percentage impedance)

Negative-sequence impedance:

$Z_2 = Z\%$ (transformer percentage impedance)

Zero-sequence impedance:

Refer to Fig. 16.10.

Cable sequence impedances are described here.

Positive-sequence impedance:

Z_1 can be calculated or determined by testing.

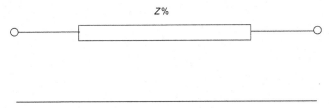

Power transformer representation in networks

FIGURE 16.5 Transformer percentage impedance.

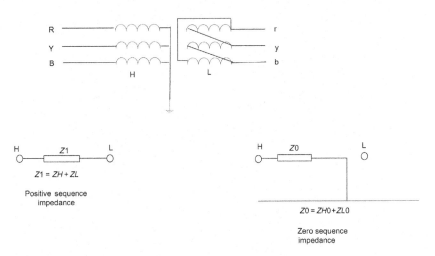

Two-winding transformer sequence impedances

FIGURE 16.6 Two-winding transformer sequence impedances.

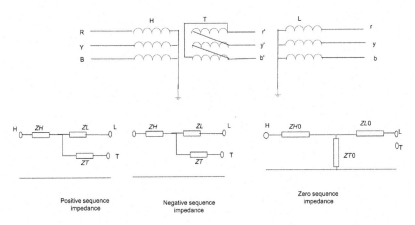

Three-winding transformer sequence impedances

FIGURE 16.7 Three-winding transformer sequence impedances.

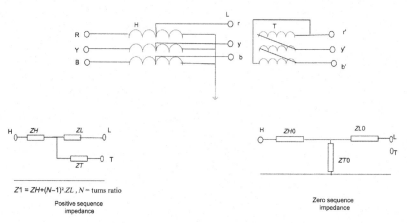

$Z1 = ZH+(N-1)^2.ZL$, N = turns ratio

Positive sequence impedance

Zero sequence impedance

Autotransformer with tertiary winding sequence impedances

FIGURE 16.8 Autotransformer with tertiary winding sequence networks.

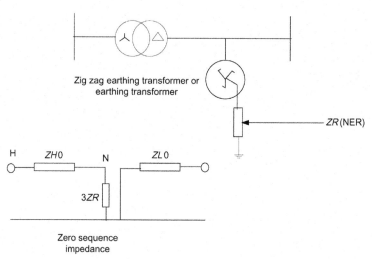

Zero sequence impedance

Note: no positive sequence impedance is there

FIGURE 16.9 Earthing transformer sequence networks.

Negative-sequence impedance:

$$Z_2 = Z_1$$

Zero-sequence impedance:

$$Z_0 > Z_1$$

Motor sequence impedances are described here.

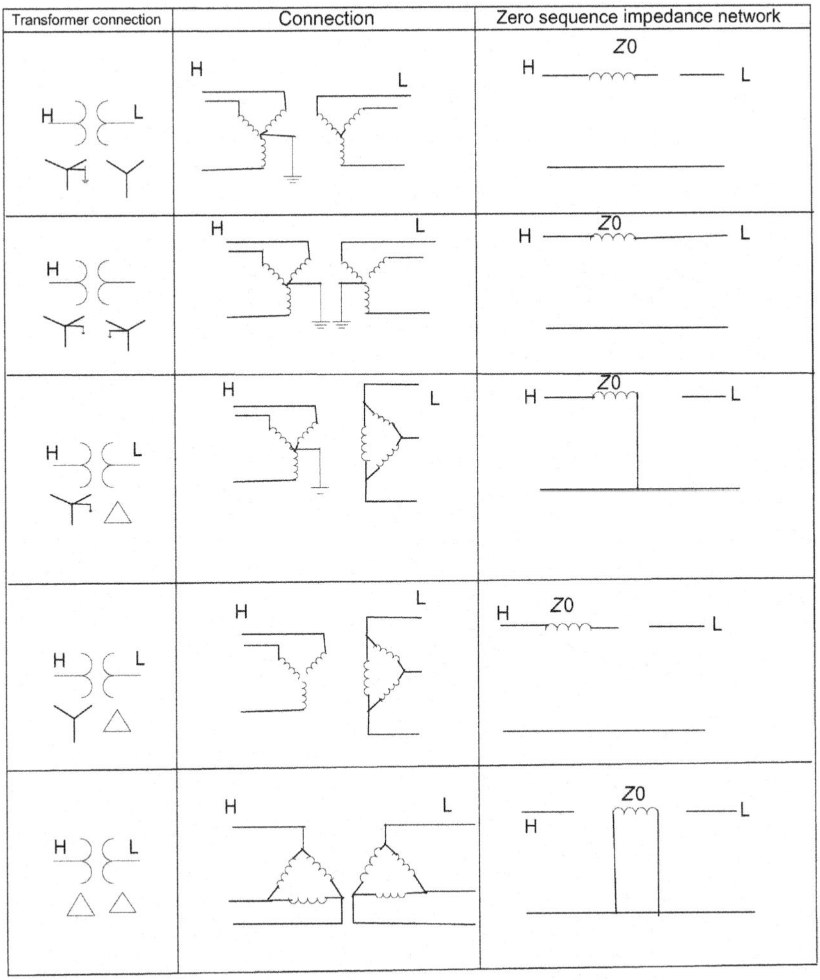

Transformer connection	Connection	Zero sequence impedance network

FIGURE 16.10   Zero-sequence impedance networks of two-winding transformers.

Positive-sequence impedance:
$X1$ = subtransient reactance
Negative-sequence impedance:

$$X_2 = X_1$$

Zero-sequence impedance:

$$Z_0 = 0$$

Transmission line sequence impedances are described here.
Positive-sequence impedance:
Z_1 can be calculated

Negative-sequence impedance:

$$Z_2 = Z_1$$

Zero-sequence impedance:

Z_0 can be calculated or determined by testing.

Synchronous generator sequence impedances are now described.

The positive-sequence impedance of a generator is approximately equal to: Xs (steady state reactance).

The negative-sequence impedance of a generator is approximately equal to: Xs'' (sub transient reactance).

The zero-sequence impedance of a generator is approximately equal to: XL (leakage reactance).

16.4 SYMMETRICAL THREE-PHASE FAULT ANALYSES

This type of fault is a symmetrical fault as it is on the three phases at the same time as shown below (Fig. 16.11):

$$V_R = V_Y = V_B, \quad I_R + I_Y + I_B = 0$$

$$I_1 = \frac{E}{Z_s + Z_1}$$

16.4.1 Unsymmetrical Single-Phase Fault Analyses

This fault is unsymmetrical as it is on one phase and earth, and can be analyzed as follows (Fig. 16.12):

$$V_R = 0, \quad I_Y + I_B = 0, \quad I_1 = I_2 = I_0, \quad I_R = 3I_1$$

$$I_1 = \frac{E}{3Z_s + 2Z_1 + Z_0}$$

$$I_F = 3I_0$$

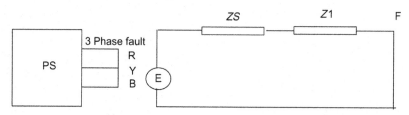

FIGURE 16.11 Three-phase fault.

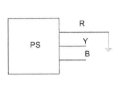
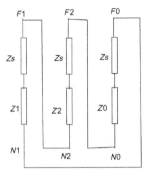

FIGURE 16.12 Single phase to earth fault.

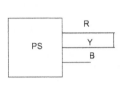
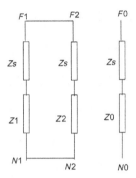

FIGURE 16.13 Phase-to-phase fault.

16.4.2 Unsymmetrical Phase-to-Phase Fault Analyses

This fault is unsymmetrical as it is from one phase to another phase and it can be analyzed as follows (Fig. 16.13):

$$V_R = V_Y, \quad I_R = -I_Y, \quad I_1 = -I_2, \quad I_B = 0$$

$$I_1 = \frac{E}{2Z_s + 2Z_1}$$

16.4.3 Unsymmetrical Phase-to-Phase Faults to Earth Analyses

This fault is unsymmetrical as it is on one phase to another phase to ground and can be analyzed as follows (Fig. 16.14):

$$I_2 + I_0 = -I_1$$

$$I_1 = \frac{E}{Z_1 + Z_s + \left(\frac{(Z_s + Z_2)(Z_s + Z_0)}{2Z_s + Z_2 + Z_0} \right)}$$

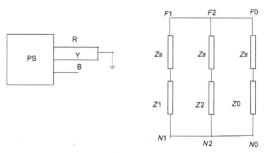

FIGURE 16.14 Phase to phase to earth fault.

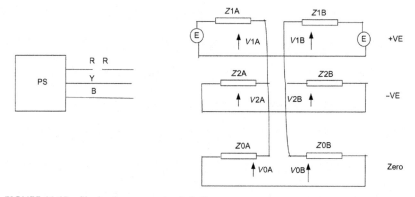

FIGURE 16.15 Single-phase open-circuit fault.

16.4.4 Unsymmetrical Single-Phase Open-Circuit Fault Analyses

This fault is unsymmetrical as it is on one phase only and can be analyzed as follows (Fig. 16.15):

$$I_2 + I_0 = -I_1$$

16.4.5 Three-Phase Equivalent Fault Level and Fault Current

To make the calculation simple, a single-phase fault can be represented approximately as three-phase faults, for example:

$$\text{MVA } 3\Phi \text{ equivalent to } 1\Phi \text{ fault} = 3\left[\frac{\text{MVA}_b \times 100}{2Z_1 + Z_0}\right]$$

$$I_F = \frac{\text{MVA}3\Phi\text{equivalent}}{\sqrt{3}V_L}$$

Sequence voltages for different fault types.

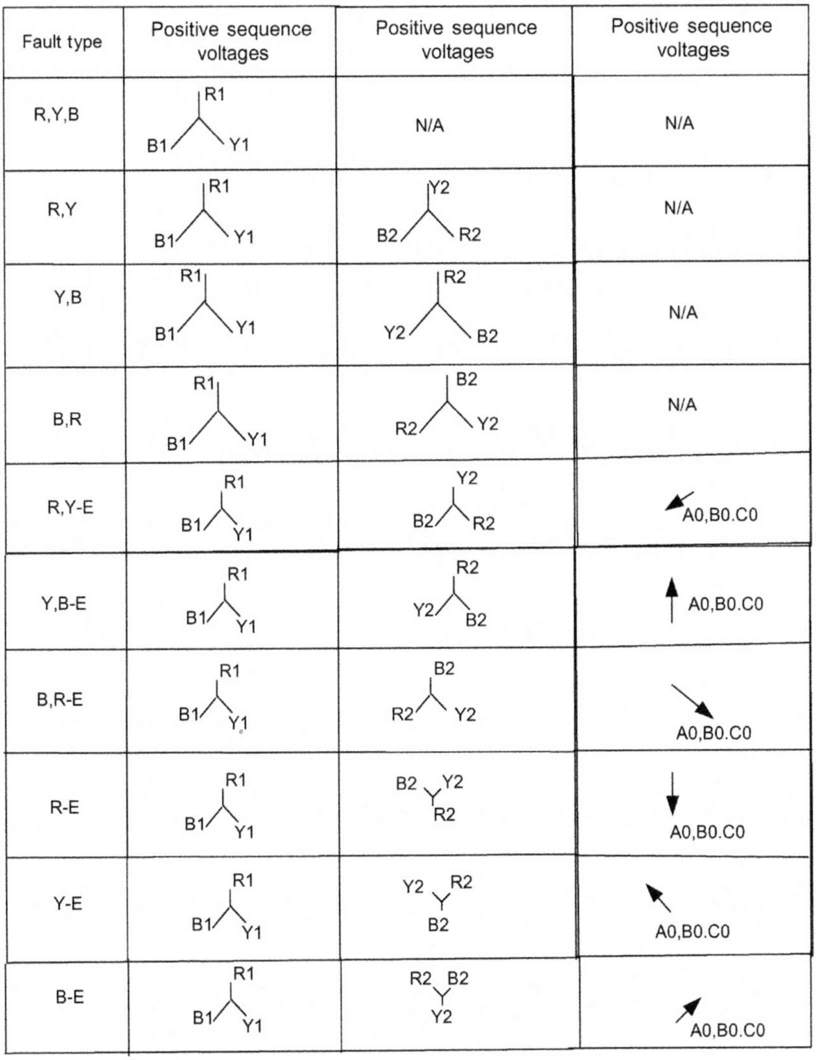

Fault type	Positive sequence voltages	Positive sequence voltages	Positive sequence voltages
R,Y,B	R1, B1, Y1	N/A	N/A
R,Y	R1, B1, Y1	Y2, B2, R2	N/A
Y,B	R1, B1, Y1	R2, Y2, B2	N/A
B,R	R1, B1, Y1	B2, R2, Y2	N/A
R,Y-E	R1, B1, Y1	Y2, B2, R2	A0,B0.C0
Y,B-E	R1, B1, Y1	R2, Y2, B2	A0,B0.C0
B,R-E	R1, B1, Y1	B2, R2, Y2	A0,B0.C0
R-E	R1, B1, Y1	B2, Y2, R2	A0,B0.C0
Y-E	R1, B1, Y1	Y2, R2, B2	A0,B0.C0
B-E	R1, B1, Y1	R2, B2, Y2	A0,B0.C0

FIGURE 16.16 The different sequence voltage phasors for different fault types.

Refer to Fig. 16.16 for the different sequence voltage phasors for different fault types.

16.5 WORKED EXAMPLES

1. *Impedance.*

Impedance on 100 MVA base = (impedance on rating) $\times$ (100/rated MVA)

For example, a 240 MVA SGT has a name plate rating of 20%;

$$\text{Rating on 100 MVA} = \frac{20 \times 100}{240}$$

$$= \mathbf{8.33\%}$$

2. *MVA*

Fault rating and fault in-feed are expressed in MVA. This is related to percent impedance by the following:

Percent impedance $= 10,000/\text{MVA}$ if the base is 100 MVA

For example, if the fault infeed at a particular substation is 35,000 MVA, then the source impedance, expressed as a percentage on 100 MVA base, will be 0.286.

Similarly, if it is calculated that the impedance to a particular fault is 2% (on 100 MVA), then the fault infeed will be 5000 MVA.

To derive the fault current (I_f), either:

a. Use $I_f = \text{fault infeed (MVA)}/(kV \times 1.732)$ kA
b. Or $I_f = 10,000/(kV \times 1.732 \times Z\%)$ kA

3. Ohmic impedance

To convert percent impedance to primary ohmic impedance:

$$\text{Taking } Z\% = \text{percent impedance on 100 MVA}$$

$$Z\% = \frac{10^4}{\text{MVA}}$$

$$\text{MVA} = \frac{V2}{Z}$$

$$Z\% = \frac{10^4 \times Z}{V^2}$$

Then, primary impedance $(\Omega) = Z\% \times (kV^2/10,000)$
For example: if $Z\% = 2$, and $kV = 400$, then $Z_{\text{prim}} = 32 \ \Omega$.
Conversely, $Z\% = Z_{\text{prim}} (10,000/kV^2)$.

To convert primary ohmic impedance to secondary ohmic impedance, this can be done by using the following impedance conversion factor
This depends upon the VT and CT ratios, as follows:

$$Z_{\text{sec}} = Z_{\text{prim}} \times (\text{CT ratio}/\text{VT ratio})$$

For example, for 400 kV, with 2000/1 CTs:

$$Z_{\text{sec}} = Z_{\text{prim}} \times (2000 \times 110/396,000)$$

General notes for calculating the impedance to fault and fault current are described here. To calculate the fault current, it is first necessary to draw out the network involved. From this the impedance network is constructed, starting from an infinite source and leading to the fault position. All values of impedance used to construct the network must refer to the same MVA base. For consistency, always use the percent impedance to a base of 100 MVA.

If the fault being considered is a three-phase fault, then there will only be positive-sequence quantities present, therefore only positive-sequence impedances need to be considered.

The impedance to the fault can now be calculated, and from this the fault current can then be calculated (I_F). The fault current in other parts of the network can be derived using network analysis techniques, with the fault current at the relay point.

If the protection relay is an impedance measuring type (i.e., distance protection) then it may only be necessary to calculate the impedance to the fault.

Example 1:
Three-phase balanced fault (Fig. 16.17)
Calculations:

Total impedance $= Z = 0.4 + 1.5 + 8.3 = 10.2$
$\text{MVAF} = 10{,}000/Z\% = 10{,}000/10.2 = 980 \text{ MVA}$
$I_F = \text{MVAF}/(kV \times 1.732) = 980/(132 \times 1.732) = 4.286 \text{ kA}$

Example 2:
Single phase-to-earth fault (Fig. 16.18)

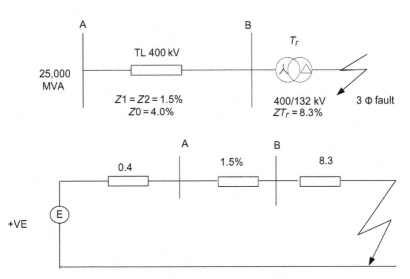

FIGURE 16.17   Three-phase balanced fault.

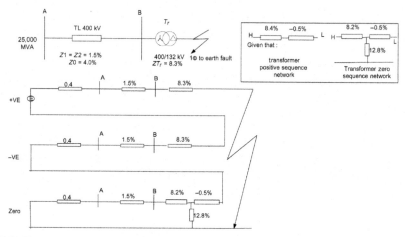

FIGURE 16.18 Single phase to earth fault.

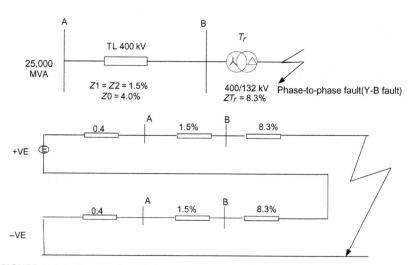

FIGURE 16.19 Phase-to-phase fault.

Calculations:

$Z_{total} = 10.2 + (10.2 + ((0.4 + 4 + 8.2) \times 12.8)/(0.4 + 4 + 8.2 + 12.8))$
$\quad\quad = 26.25\%$
$I_1 = I_2 = I_0 = 10{,}000/(Z\%) = 10{,}000/26.25 = 381 \text{ MVA}$
$I_F = I_1 + I_2 + I_0 = 3 \times 381 = 1143 \text{ MVA}$
$I_F = \text{MVAF}/kV \times 1.732 = 1143/(132 \times 1.732) = 4.999 \text{ kA}$

Example 3:
Phase-to-phase fault (Fig. 16.19)
Calculations:

$Z_{total} = 10.2 + 10.2 = 20.4\%$
$MVAF = 10,000/20.4 = 490$ MVA
$I_F = I_1 = -I_2 = 490$ MVA
$I_Y = -I_B = 1.732 \times 490 = 849$ MVA

Chapter 17

IEC 61850 Protocols Used in Protective Relays Communication

17.1 INTRODUCTION

In 2005, a new standard, IEC 61850, that defined communication protocols for intelligent electronic devices (IEDs) at electrical substations was introduced. It discusses how substations and power systems components communications should be designed, built, commissioned, operated, and maintained.

The same technology is used in transmission and distribution systems.

The substation automation system (SAS) was introduced in 1980 in transmission and distribution substations. Once IEDs were introduced in protection and control, local area network (LAN) architecture was used to link the IEDs.

However, communication is only possible between the linked devices with the use of protocol conversions. The speed of the IED protocols limited the functionality and services and varied between manufacturers. Once the IEC 61850 protocol for communication networks and systems in substations was introduced, it allowed for standardized high speed communication between IEDs from different manufactures using the protocol. This meant that companies could reduce the use of expensive standalone devices and complex real copper wiring and convert to a new protocol that allows engineers fully use the functions of IEDs in protection and control of substations.

Table 17.1 shows a list of common abbreviations used in this field.

17.2 IEC 61850

The use of IEC 61850 is to help standardize the communication within substations with low costs and increased flexibility for all functions between different IEDs. It is a structured definition of data within devices over any communication media and protocol used.

The use of this standard reduces installation and maintenance costs by reducing manual configuration and hard wire costs. More functions are available in IEDs that meet this standard, giving more flexibility to exchange

Practical Power System and Protective Relays Commissioning.
DOI: https://doi.org/10.1016/B978-0-12-816858-5.00017-4

TABLE 17.1 Common Abbreviations Used in the Communications Protocol Field

Acronym	Meaning
ABSI	Abstract communication service interface
ASDU	Application service data unit
BRCB	Buffered report control block
CDC	Common data class
CT	Current transducer
DTD	Document type definition
DUT	Device under test
FAT	Factory acceptance test
GI	General interrogation
GoCB	GOOSE control block
GOOSE	Generic object oriented substation events
GSE	Generic substation event
HMI	Human machine interface
ICD	IED capability description
IED	Intelligent electronic device
IP	Internet protocol
LCB	Log control block
LD	Logical device
LN	Logical node
MC	Multicast
MCAA	Multicast application association
MICS	Model implementation conformance statement
MMS	Manufacturing message specification (ISO 9506 series)
MSVCB	Multicast sampled value control block
PICS	Protocol implementation conformance statement
PIXIT	Protocol implementation eXtra information for testing
PUAS	Power utility automation system
RTU	Remote terminal unit
SAT	Site acceptance test

(Continued)

TABLE 17.1 (Continued)

SAV	Sampled analog value (IEC 61850-9 series)
SCADA	Supervisory control and data acquisition
SCD	Substation configuration description
SCL	Substation configuration language
SCSM	Specific communication service mapping
SGCB	Setting group control block
SoE	Sequence-of-events
SSD	System specification description
SUT	System under test
SV	Sampled values
SVCB	Sampled value control block
TCP	Transport control protocol
TPAA	Two party application association
TUT	Tool under test
URCB	Unbuffered report control block
USVCB	Unicast sampled value control block
UTC	Coordinated universal time
VT	Voltage transducer
XML	eXtensible markup language

information between IEDs produced by different manufacturer. In other words it is a one common language between the control, protection, and monitoring devices.

17.2.1   Fundamental Design and Operation

In IEC 61850, functions within substations are termed as follows:

- "Station level" refers to the substation as a whole.
- "Bay level" refers to the feeder, circuit breaker (CB) or other substation components.
- "Process level" refers to the high-voltage power system or primary equipment and sensors.

  This hierarchy is shown in Fig. 17.1.

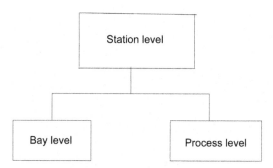

FIGURE 17.1   Functions within substations as reflected in IEC 61850.

17.2.1.1   Logical Nodes and Services

The standard data consists of different logical nodes; the access to this data is called the service method. There are logical nodes for automatic control, the names of which all begin with the letter "A." There are also logical nodes for metering and measurement, the names of which all begin with the letter "M." Likewise there are logical nodes for supervisory control (C), generic functions (G), interfacing/archiving (I), system logical nodes (L), protection (P), protection related (R), sensors (S), instrument transformers (T), switchgear (X), power transformers (Y), and other equipment (Z).

A CB is modeled as an XCBR logical node. It contains a variety of data including "Loc" for determining if operation is remote or local, "OpCnt" for an operations count, "Pos" for the position, "BlkOpn" for block breaker open commands, "BlkCls" for block breaker close commands, and "CBOpCap" for the CB operating capability.

Each element of data within the logical node conforms to the specification of a common data class (CDC) per IEC 61850-7-3. Each CDC describes the type and structure of the data within the logical node.

For example:

Relay1 XCBR1 ST Loc stVal represents the following:

Relay1 = logical device
XCBR1 = logical node
ST = functional constraints
Loc = data
stVal = attribute

17.2.1.2   Physical Devices

The LAN is used to communicate between physical nodes. There are two interfaces: physical interfaces between physical nodes, and logical interfaces between logical nodes.

17.2.1.3   Communication in IEC 61850

The services and protocols used for communicating under IEC 61850 are:

- Sampled value service
- Goose service (GOOSE)
- Simple network time protocol
- Manufacturer message specification (MMS)
- User datagram protocol
- Transport control protocol
- Internet protocol

Each physical device can be a "publisher", that is, create information, and send it using abstract communication services interface (ACSI) communication to other physical devices that are nominated as "subscribers." These ACSI communications can be in the form of GOOSE messages or generic substation event (GSE).

MMS defines the communication messages between controllers and between engineering stations and controllers (downloading and application or reading/writing variables).

17 .2.1.4   Transport Profile

The above are the different services used for communication in IEC 61850.

The data and services of an application can be classified in the following levels:

- Abstract communication service interface (ACSI) the mechanism of communication services.
- Common data class.
- Compatible addressing of logical nodes and data objects.

17.2.1.5   Bus Topology

This can be one of the following configurations:

- Single and double star topology
- Single and double tree topology
- Mesh topology
- Ring topology

17.2.2   Substation Configuration Language

IEC 61850-6-1 specifies a substation configuration language (SCL) that is based upon eXtensible Markup Language (XML) to describe a system configuration. SCL specifies a hierarchy of configuration files that enable multiple levels of the system to be described in standardized XML files. The various SCL files include system specification description (SSD), IED capability description (ICD), substation configuration description (SCD), and configured IED description files. All these files are constructed in the same methods and format but have different scopes depending upon the need.

17.2.2.1 Setting up Protocols Using IEC 61850

The first step is to define the SAS functions, IEDs and their relation to the primary substation components. Additional notes should be made on any specific functions and the information to be transferred to other IEDs in the system. These functions should be defined using SCL.

An IED-ICD file is generated and transferred to SCD file by an IED configuration tool; this is done for each different IED in the system.

Additionally, the system specification can be described in SCL by a SSD file and/or transferred to a SCD file using a system specification tool.

Then the SCD-61850 file is downloaded to each IED using the system configuration tool.

The SCD file defines the substations, IEDs, communication, and logic nodes using SCL.

The following items should be confirmed by the manufacturer to confirm compatibility with IEC-61850:

- Protocol implementation conformance statement.
- Protocol implementation eXtra information for testing.
- Model implementation conformance statement.

17.2.2.2 Security of IEC-61850

Using IEC-61850 retains the security of the following criteria:

- Confidentiality: protecting information.
- Authentication: clear definition of user.
- Access control: different level of access.
- Message integrity: ensures that message is not tampered with.

17.3 GENERAL DESCRIPTION OF A SUBSTATION AUTOMATION SYSTEM

As discussed earlier in this chapter, an automatic control system in a substation are divided in three levels locally in substation, and one level in the master station, as follows:

1. Master station level.
2. Substation level.
3. Bay level.
4. Process level.

A substation automation system such as that shown in Fig. 17.2 has different interface systems, hardwired or optical interfaces, and a process bus.

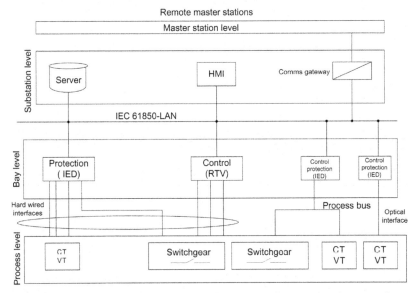

FIGURE 17.2 Substation automation setup. *CT*, current transformer; *VT*, voltage transformer; *LAN*, local area network.

17.4 TESTING AND COMMISSIONING OF IEC 61850 IN SUBSTATION

The benefits of using IEC 61850-compatible devices includes lower commissioning costs, and decreased configuration and commissioning costs as the devices do not need as much manual configuration as legacy devices can require. Client applications no longer need to manually configure for each point they need to access because they can retrieve the points list directly from the device or import it via an SCL file. Many applications require nothing more than setting up a network address in order to establish communications. Most manual configuration is eliminated, drastically reducing errors and rework.

17.4.1 IEC-61850 Testing

Testing procedures include the following:

- Functional element testing
- Integration testing
- Function testing
- System testing

Tests also can be classified as follows:

- Factory acceptance testing
- Site acceptance testing
- Complete system integration testing.

SCL and IEC 61850 define the substation/system configuration description testing.

This is done using the following tools:

- Diagnostic tool: these isolate the cause of the problem which is detected by self-monitoring supervision of the system or the problem detected during testing process.
- Simulation tool: these simulate the conditions and signals to test the system.
- Documentation tool: reports the results in a SCD file.

In general the commissioning procedures for the protection and control systems based on IEC 61850 can be done as follows:

1. The tests which can be classified as secondary tests in conventional system:
   a. Relay (IED) testing as normal conventional system.
   b. Relay isolation (blocking of relay outputs for trip signals) by a manual command that puts the relay into test isolated mode then all commands from the IED (relay) going through the process bus will be disabled. At that moment the relay screen will show the LED in service in red or "off" position.
   c. Connect the relay injecting set to the substitute merging units after disconnecting the original merging units which connect the primary equipment in switchgear to the relay. Put the relay in test-substituted mode to test the relay through its input fibers and substitute merging units.
   d. Relay restoration to service: reconnect the fiber cables of the original merging units of the relay and put the relay into "In service mode."
2. The tests which can be classified as primary tests in conventional systems/merging unit/primary equipment:
   a. Use a temporary relay and inject the primary equipment with primary or secondary injection and note the values measured by the temporary relay.
   b. Issue the commands from the temporary relay to the CB and check the operation of the CB in the switchgear.
   c. Note any alarms or signals coming from the CB to the temporary relay.

See also Fig. 17.3.

3. Final test of SAS (Substation Automation System)
   a. Test the total automation system of substation by issue the manual tripping orders and closing orders to the CB through human machine interface to the substation automatic system.

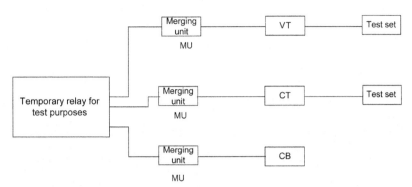

FIGURE 17.3 Primary equipment test with merging unit. *CT*, current transformer; *VT*, voltage transformer; *CB*, circuit breaker.

  b. Test the operation of the equipment in the primary system, switchgear, disconnectors and earthing switches.
  c. During these tests any alarms should be noted, and an inspection of the status of each device in the mimic diagram of substation topology.
  d. Check the operation of all protection functions and relevant alarms and automatic tripping orders.

Chapter 18

Protection Relays

Subchapter 18.1

Introduction

Relays are devices connected throughout a power system to detect any unbalance or abnormal condition. There are many types of relays used to protect power systems and these are described in this chapter.

Subchapter 18.2

Classification of Relays

18.2.1 CLASSIFICATION BASED ON FUNCTION

1. Protective relays are relays that detect abnormal conditions in the system and issue an alarm or trip based on the severity of the fault.
2. Monitoring relays are relays that monitor the system but do not issue any trip signal, such as synchronism verification relays.
3. Programming relays have electrical sequences, such as autoreclosing and synchronizing relays.
4. Regulating relays, such as voltage regulator relays, regulate the transformer voltage by changing the on load tap changer.
5. Auxiliary relays include repeater relays and trip lookout relays.

18.2.2 CLASSIFICATION BASED ON INPUT QUANTITIES

1. Current relays;
2. Voltage relays;
3. Power relays;
4. Pressure relays;

Practical Power System and Protective Relays Commissioning.
DOI: https://doi.org/10.1016/B978-0-12-816858-5.00018-6

5. Frequency relays;
6. Temperature relays;
7. Flow relays;
8. Vibration relays.

18.2.3 CLASSIFICATION BASED ON PRINCIPALS OR STRUCTURES

1. Differential relays;
2. Impedance relays;
3. Electromechanical;
4. Solid state-static relays;
5. Digital-static relays;
6. Numerical-static relays;
7. Thermal relays.

18.2.4 CLASSIFICATION BASED ON CHARACTERISTICS

1. Distance relays;
2. Directional relays;
3. Inverse time relays;
4. Definite time (DT) relays;
5. Under voltage relays;
6. Earth fault or phase fault relays;
7. High- or low-speed relays;
8. Directional comparison relays;
9. Segregated phase relays;
10. High-impedance or low-impedance differential relays;

Subchapter 18.3

Design of Protective Relaying Systems

The design of a protection power system is divided into several zones; each zone needs a group of relays. The design of a protective relaying system depends on the following factors:

1. Economics, the importance of the protected power system components and the required degree of protection against the cost.
2. Availability of the measured input signals, current transformers (CT), and voltage transformer (VT) locations.
3. Design philosophy used before, and previous experience.

18.3.1 DESIGN CRITERIA OF PROTECTIVE RELAYING SYSTEMS

1. Reliability

The ability of the protective system to perform correctly when needed (dependability) and to avoid unnecessary operation (security).

2. Speed

This is the operating time of the relay, which is a very important factor to reduce the fault damage effects on the system.

3. Selectivity

This is the maximum service continuity with minimum system disconnection and is done by tripping the only faulted part of the system and leaving the healthy part of the system in operation (refer to Fig. 18.3.1).

Where GS = generator, T_1 = step-up transformer, T_2 = step-down transformer, TL = transmission line, Zone 1, Zone 2, and Zone 3 are protection zones, R_1 = relay in zone 1, R_2 = relay in zone 2, A1, A2, A3 are busbars of substations.

As shown in Fig. 18.3.1 the selectivity means that the relay R_2 should detect the fault F on the transmission line (TL) and isolate this fault instantaneously. Selectivity also means that the relay R_1 will not operate instantaneously on the fault F but will operate on the fault F with a time delay when the relay R_2 does not operate instantaneously.

4. Security

Security means that the relay should not operate for external faults of the protected zone or operate for heavy load conditions.

5. Simplicity

This means the use of minimum equipment and maximum security.

6. Economics

This means maximum protection with minimum costs.

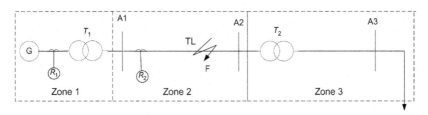

FIGURE 18.3.1 Power system protection zones' selectivity and sensitivity.

7. Dependability

This means that the relay should operate for faults in its zone in the correct way and at the correct time, with selectivity, sensitivity, and speed, to give a proper trip for the fault in its zone.

8. Sensitivity

Sensitivity is critical to being able to detect faults. There are three independent components for the measured input quantities for the relay: accuracy, precision, and resolution.

18.3.2 ZONES OF PROTECTION

The general philosophy of protection is to divide the power system into protective zones that can be protected adequately with a minimum number of disconnections.

These zones are overlapped to avoid any possibility of unprotected areas in the power system and can be divided into the following protective zones:

Generator zone, Zone B in Fig. 18.3.2;
Transformer zone, Zone A in Fig. 18.3.2;
Busbars zone, Zone C in Fig. 18.3.2;
Transmission and distribution circuits zone, Zones D and E in Fig. 18.3.2;
Motors zone, Zone F in Fig. 18.3.2.

Fig. 18.3.2 shows the protection zones and their overlapping.

18.3.3 STEPS AND INFORMATION REQUIRED TO DESIGN A GOOD PROTECTION SYSTEM

1. System configuration based on a single line diagram (SLD).
2. Existing system protection based on a protection SLD (PSLD).
3. Existing operation procedures, practices, and possible extensions.
4. Fault study, with the maximum and minimum fault levels during summer and winter.
5. Maximum load, CT ratios, VT ratios.
6. System parameters, such as transmission line impedances, transformer impedances.
7. CT and VT locations, connections, and ratios.

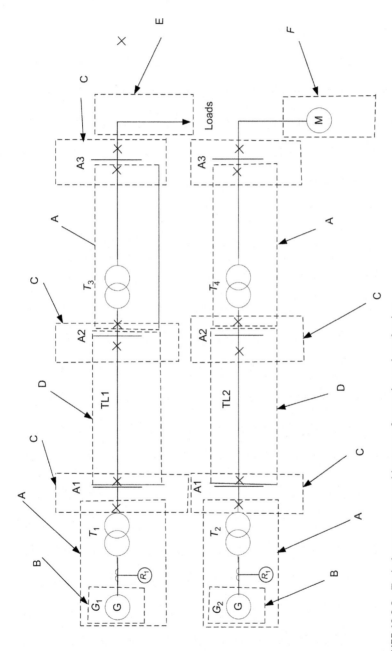

FIGURE 18.3.2 Typical power system and its zones of protection and overlapping.

Subchapter 18.4

History of Relays

Relays began as electromechanical relays, which employ the magnetic coils, then developed to static relays, which employ transistors, then digital relays, which employ microprocessors, and finally numerical relays, which use digital relays, which can communicate with each other with different control protocols.

Different types of relays include the following:

1. Electromechanical relays;
2. Static relays—analog relays;
3. Static relays—digital relays;
4. Static relays—numerical relays.

18.4.1 PRINCIPLES OF THE CONSTRUCTION AND OPERATION OF THE ELECTROMECHANICAL IDMT RELAY

These relays, for example, an induction disk, have an inverse time characteristic with respect to the current input to the relay from the secondary of the CT, and these relays were used in the past and are still sometimes used in 11-kV switchgear, generator protection, and 66-kV line protections. The same principles are used in static and digital recent relays (see Fig. 18.4.3).

The above relay is electrically represented as shown in Fig. 18.4.4.

The force that rotates the disk is proportional to the product of I1 × I2 sinα, where α is the angle between I1 and I2, and this relay has an inverse characteristic as shown in Fig. 18.4.5.

It can be seen that the operating time of an Inverse Definite Minimum Time (IDMT) relay is inversely proportional to a function of the current, that is, it has a long operating time at low multiples of setting current and a relatively short operating time at high multiples of setting current.

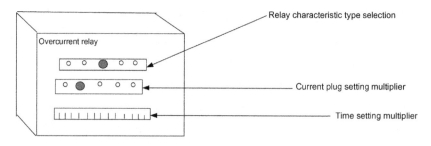

Electromechanical relay

FIGURE 18.4.3 Electromechanical relay—overcurrent type.

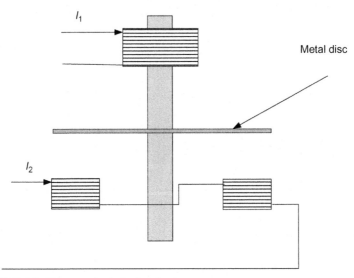

FIGURE 18.4.4   Electromechanical relay circuit.

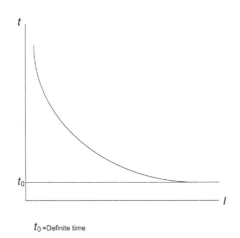

t_0 =Definite time

FIGURE 18.4.5   Relay inverse characteristic.

The characteristic curve is defined by BS 142 and, as shown in Fig. 18.4.5, has two adjustments on the relay:

1. The current plug setting multiplier (PSM), which can be set based on fault current as a multiplier of the setting current;
2. Time setting multiplier (TMS), which reflects the operating time of the relay as a multiplication of the time multiplier (TM) by the relay operating time (TR) based on the relay characteristic type chosen on the relay, for example, the standard inverse (SI) characteristic.

At the end of the characteristic at a certain multiplication of the current the time becomes DT (t_0).

There are different types of characteristic which we can used based on the application of the relay, as shown in Fig. 18.4.6.

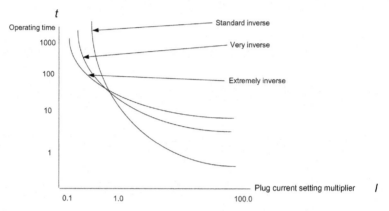

FIGURE 18.4.6 Some types of inverse definite minimum time characteristics.

Based on the relay setting calculation we can choose the TM from the relay characteristic curve at the PSM where we can find the relay operating time as shown in Fig. 18.4.7.

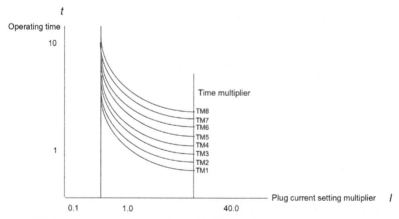

FIGURE 18.4.7 Time/current characteristic with different time multiplier.

The relay burden is the continuous load on the CT circuit by the relay and circuit leads and is expressed by V_A or ohmic resistance.

18.4.2 STATIC PROTECTION RELAY

Static protection relays are the electronic relays which have no moving parts, these relays can be analog transistor types in electric circuits or digital microprocessor relays. The burden of these relays is very low and in recent digital relays, the relay can perform multiple functions based on the operating algorithm. Also, by using these relays we can generate relay characteristics which cannot be generated by an electromechanical one. The cost is also reduced, and the relay can monitor its internal circuits and give an alarm if something is wrong and can block its operation if there is an internal fault which cannot be solved by the relay itself (the relay failure monitoring feature which is not available in the old electromechanical type), the maintenance is also reduced as there are no moving parts in these relays.

On the other hand, these analog relays perform with many failures at the beginning of its used when used at site some faults in relay cards and in its DC supply converter but it is solved later and this is very rare in digital relays, refer to Figs. 18.4.8 and 18.4.9 for static analog relay and digital relay, respectively.

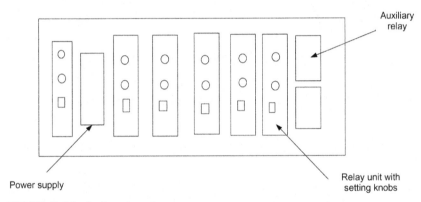

FIGURE 18.4.8 Static analog relays.

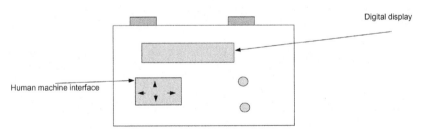

FIGURE 18.4.9 Static digital relays.

18.4.3 INTELLIGENT ELECTRONIC DEVICE RELAYS

IEDs refer to intelligent electronic devices, which are multifunctional digital relays with a built-in algorithm and can communicate with other IEDs in the substation and with the substation control system. These IEDs can carry out control functions and replace the hardwire interlocking system, circuit breaker (CB) local control, and monitoring of SF_6 in a gas insulated system (GIS). They can replace the local control cubicle with the bay controller IED and can also be a protection device which can perform as, for example, a distance protection or busbar protection. They are marketed by a number of different manufacturers. The most important advantage of this type of relay is that it can be used in advanced adaptive protection functions, where one relay can communicate with another relay through a communication link protocol to take the trip decision by the operation of the two relays.

18.4.3.1   Protection Intelligent Electronic Device

Most manufacturers classify it by the feeder circuit, which is protected by this type of IED, for example, as follows.

The feeder protection IED includes the following functions and may include more functions based on the system requirements:

Overcurrent protection—directional and nondirectional;
Distance protection;
Line differential protection with a fiber-optic link between the two sides of the line;
Autoreclosure function;
Synchro check function.

The generator protection IED includes the following functions and may include more functions based on the system requirements:

Overcurrent protection voltage-restrained or voltage-controlled;
100% stator earth fault protection;
95% stator earth fault protection;
Differential protection;
Restricted earth fault protection;
Back-up distance protection;
Reverse power protection;
Negative-phase sequence current protection;
Loss of excitation protection;
Out-of-step protection;
Over/under voltage protection;
Over/underfrequency protection.

The transformer protection IED includes the following functions and may include more functions based on the system requirements:

Overload thermal replica impedance protection;
Overcurrent protection in two stages;
Earth fault protection in two stages;
Differential protection;

Restricted earth fault protection;
Binary inputs for transformer mechanical protection, such as Buchholz protection, safety value, winding and oil temperature protection, and cooling fan system automatic operation;
Overfluxing (voltage/frequency ratio) protection.

The busbar protection IED includes the following functions and may include more functions based on the system requirements:

High-impedance busbar differential protection or low-impedance biased differential protection—it can be of the centralized or decentralized type;
Breaker failure (BF) protection;
Relay failure protection.

As we see above, we note that one IED can perform many protection functions which were previously performed by a separate relay for each function. This means more cost saving and more accuracy with the built-in algorithm inside the IED.

Fig. 18.4.10 illustrates a protection IED relay.

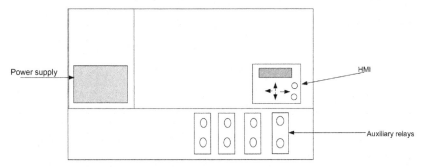

FIGURE 18.4.10   Protection intelligent electronic device relay.

18.4.3.2   Control and Monitoring Intelligent Electronic Device

Control and monitoring of IEDs include the following functions and may include more functions based on the system requirements:

Local control switching of CBs, isolators, and earthing switches;
Metering measurements including, P, Q, $\cos \Phi$, I and V, and F;
Event recording;
Circuit interlocking which in practice is done using hard wires and in parallel software interlocking inside the IED;
Trip circuit supervision, but some utilities use a separate relay for this function—an old electromechanical one;
SF_6 gas monitoring in GIS high-voltage system;
Event and sequence recording.

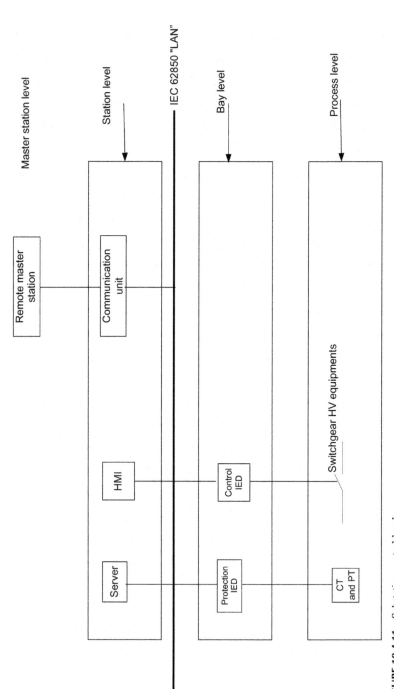

FIGURE 18.4.11 Substation control levels.

18.4.3.3   Substation Control System

In the 1980s, a substation automation system was introduced to the transmission and distribution sector in substations, making the engineering easier and also reducing the cost. The need for a communication system which can connect the different types of relays from different manufacturers led to integrate of the IEDs in local area network (LAN) architecture using a protocol called IEC 61850, which is the standard mostly common used today.

Substation automation architecture is divided into the following levels (as shown in Fig. 18.4.11):

Master station level;
Station level;
Bay level;
Process level.

Subchapter 18.5

Overcurrent Relays

18.5.1   INTRODUCTION

Overcurrent relays can be set as DT characteristic or inverse definite minimum time characteristic (IDMT), and can be applied to protect overhead lines and to be coordinated with an optimal coordination time of 0.4 second as shown in Figs. 18.5.1 and 18.5.2 to achieve selectivity and coordination by time grading with the two philosophies available, namely:

1. Definite time; or
2. Inverse definite minimum time.

For the first option, the relays are graded using a DT interval of approximately 0.5 second, an optimal 0.4 second at the relay R_3, at the extremity of the network, is set to operate in the fastest possible time, whilst its upstream relay R_2 is set 0.5 second higher. Relay operating times increase sequentially at 0.5-second intervals on each section, moving back toward the source as shown in Fig. 18.5.1.

The problem with this philosophy is that the closer the fault is to the source, the higher the fault current and the slower the clearing time—which is exactly the opposite to what we should be trying to achieve. On the other hand, inverse curves, as shown in Fig. 18.5.2, operate faster at higher fault currents and slower at lower fault currents, thereby offering the features that are desired. This explains why the IDMT philosophy has become standard practice throughout many countries over the years.

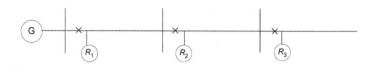

FIGURE 18.5.1 Definite time philosophy.

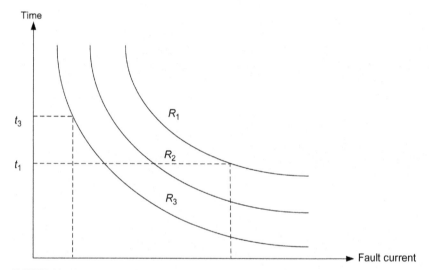

FIGURE 18.5.2 Inverse definite minimum time.

18.5.2 NONDIRECTIONAL OVERCURRENT AND EARTH FAULT PROTECTION

Relay characteristics types used in Over Current (OC) and Earth Fault (EF) relays as follows:

1. Standard inverse (SI) (normal inverse);
2. Very inverse (VI);
3. Extremely inverse (EI);
4. Long time inverse;
5. DT D2 (0.1−2.0 seconds);
6. DT D4 (0.2−4 seconds);
7. DT D8 (0.4−8.0 seconds).

The following are typical operating characteristics in use:

SI $t = \frac{0.14}{I^{0.02}-1}$

VI $t = \frac{13.5}{I-1}$

EI $t = \frac{80}{I^2-1}$

Long time standby earth fault $t = \frac{120}{I-1}$ where $t = $ relay operating time(s), $I = $ current (multiple of plug setting).

18.5.3   PRINCIPLES OF TIME/CURRENT GRADING

18.5.3.1   Discrimination by Time

In a radial system a time grading interval of 0.4 second can be used for time discrimination of faults as show in Fig. 18.5.3. The first breaker to trip for fault F is breaker D, then breaker C after 0.4 second, then breaker B after 0.4 second, then finally breaker A after 0.4 second.

$t_4 = t$, $t_3 = t + 0.4$, $t_2 = (t + 0.4 + 0.4)$, $t_1 = (t + 0.4 + 0.4 + 0.4)$.

18.5.3.2   Discrimination by Current

In this case, discrimination depends on the current and on the fact that the fault current depends on the position of the fault because of the difference in impedance values between the source and the fault, therefore relays are set in different current settings in such a way that only the relay nearest to the fault trips its breaker first, as shown in Fig. 18.5.4.

$F_1 > F_2 > F_3 > F_4$

$I_{set\ 1} > I_{set\ 2} > I_{set\ 3} > I_{set\ 4}$

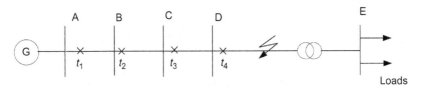

FIGURE 18.5.3   Radial power system with overcurrent time grading.

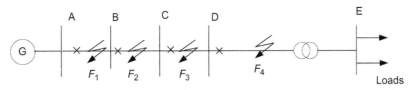

FIGURE 18.5.4   Radial power system with overcurrent current grading.

18.5.3.3 Discrimination by Current and Time

The grading can be done with both time and current.

18.5.4 EARTH FAULT PROTECTION

This protection can be done using the following CT connection in the residual path, as shown in Fig. 18.5.5.

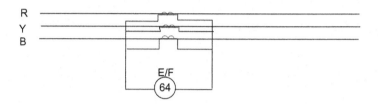

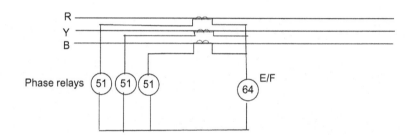

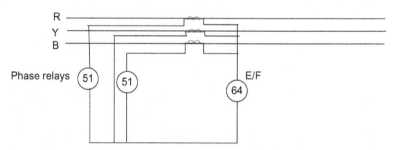

FIGURE 18.5.5 Residual connection of current transformers to earth fault relays.

18.5.5 DIRECTIONAL RELAYS

Directional relays are used to trip in one direction only by taking a voltage vector as reference to the current vector. The reference voltage is called a polarizing signal.

The angle by which the current applied to the relay is displaced from the voltage applied to the relay at the unit power factor is called the relay connection, for example, 90 degrees connection between I_R and $V_{Y\text{-}B}$.

The relay angle which produces maximum torque is the maximum torque angle (MTA).

As shown in Fig. 18.5.6, the relay connection is 90 degrees between polarizing voltage $V_{Y\text{-}B}$ and current input to the relay I_R and the MTA is 45 degrees between the maximum torque line and the reference polarizing voltage $V_{Y\text{-}B}$ (Table 18.5.1).

As shown in Fig. 18.5.7, the relay connection is 90 degrees between polarizing voltage $V_{Y\text{-}B}$ and the current input to the relay I_R and the MTA is 30 degrees between the maximum torque line and the reference polarizing voltage $V_{Y\text{-}B}$ (Table 18.5.2).

As shown in Fig. 18.5.8, the relay connection is 30 degrees between polarizing voltage $V_{R\text{-}B}$ and current input to the relay I_R and the MTA is 0 degree between the maximum torque line and reference polarizing voltage $V_{R\text{-}B}$ (Table 18.5.3).

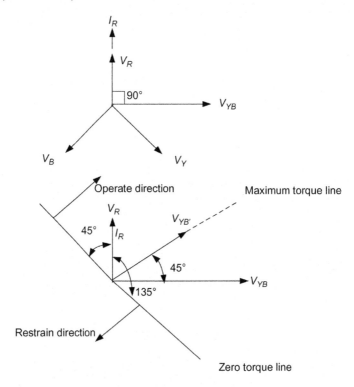

90° Connection-45° MTA

FIGURE 18.5.6   90 Degrees connection 45 degrees maximum torque angle.

TABLE 18.5.1 90 Degrees Connection 45 Degrees Maximum Torque Angle Connections

Relay	Current	Voltage
R	I_R	V_{YB}
Y	I_Y	V_{BR}
B	I_B	V_{RY}

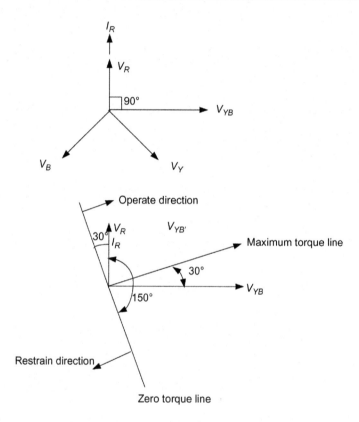

90° Connection-30° MTA

FIGURE 18.5.7   90 Degrees connection 30 degrees maximum torque angle.

TABLE 18.5.2 90 Degrees Connection 30 Degrees Maximum Torque Angle Connections

Relay	Current	Voltage
R	I_R	V_{YB}
Y	I_Y	V_{BR}
B	I_B	V_{RY}

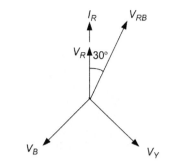

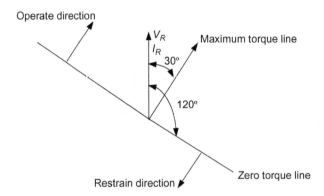

30° Connection with Zero° MTA

FIGURE 18.5.8   30 degrees connection maximum torque angle at zero degree.

TABLE 18.5.3 30 Degrees Connection Zero Degrees Maximum Torque Angle Connections

Relay	Current	Voltage
R	I_R	V_{RB}
Y	I_Y	V_{YR}
B	I_B	V_{BY}

Refer to Fig. 18.5.9 for directional relay connections and to Fig. 18.5.10 for a residual voltage connection.

The recommended MTA for directional relays is as follows:

Zero degree for resistance-earthed systems;
Forty five degrees for lag power factor distribution system solidly earthed;
Sixty degrees for lag power factor transmission systems solidly earthed.

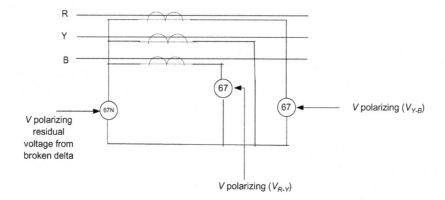

FIGURE 18.5.9 Directional relays connections.

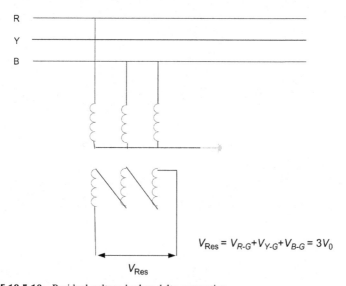

$$V_{Res} = V_{R-G} + V_{Y-G} + V_{B-G} = 3V_0$$

FIGURE 18.5.10 Residual voltage-broken delta-connection.

18.5.5.1 Ring Main Circuit Overcurrent and Directional Relays Grading

Refer to Fig. 18.5.11 for the grading method for a ring main circuit.

1. Open Ring at A
   Grade A'-E-D'-C-B'
2. Open Ring at A'
   Grade A-B-C-D-E

Grading times are given as an example in Fig. 18.5.11, relays B,C,D,E can also be nondirectional relays.

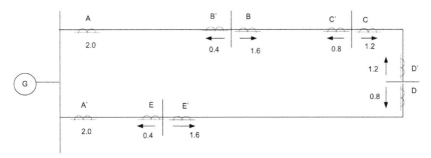

FIGURE 18.5.11 Grading method for a ring main circuit.

18.5.6 IEC CURVES AND ANCI/IEEE CURVES OVERCURRENT RELAY CHARACTERISTIC

a. IEC 60255 curves overcurrent relay characteristic (Table 18.5.4)
b. ANCI/IEEE curves overcurrent relay characteristic (Table 18.5.5)

TABLE 18.5.4 IEC 60255 Curves Overcurrent Characteristic

Relay Characteristic	Equation (IEC 60255)
Standard inverse (SI)	$t = \text{TMS} \times \dfrac{0.14}{I^{0.02-1}}$
Very inverse (VI)	$t = \text{TMS} \times \dfrac{13.5}{I-1}$
Extremely inverse (EI)	$t = \text{TMS} \times \dfrac{80}{I^{2-1}}$
Long time standard earth fault	$t = \text{TMS} \times \dfrac{120}{I-1}$

TABLE 18.5.5 ANCI/IEEE Curves Overcurrent Characteristic

Relay Characteristic	Equation (IEC 60255)
IEEE moderately inverse	$t = \dfrac{\text{TD}}{7} \left(\dfrac{0.0515}{I_r^{0.02-1}} + 0.114 \right)$
IEEE very inverse	$t = \dfrac{\text{TD}}{7} \left(\dfrac{19.61}{I_r^{2-1}} + 0.491 \right)$

(*Continued*)

TABLE 18.5.5 (Continued)

Relay Characteristic	Equation (IEC 60255)
Extremely inverse	$t = \dfrac{TD}{7}\left(\dfrac{28.2}{I_r^{2-1}} + 0.1217\right)$
US CO8 inverse	$t = \dfrac{TD}{7}\left(\dfrac{5.95}{I_r^{2-1}} + 0.18\right)$
US CO2 short time inverse	$t = \dfrac{TD}{7}\left(\dfrac{0.02394}{I_r^{0.02-1}} + 0.01694\right)$
North America IDMT relay characteristic where $I_r = I/I_s$ = relay setting current	
TMS = time setting multiplier	
TD = time dial	

Subchapter 18.6

Distance Protection

18.6.1 BASIC PRINCIPLES

Distance protection, as its name implies, means that the relay will detect the faults in the power system transmission line within a specified distance—the impedance of the transmission line is a function of the line length or distance as shown in Fig. 18.6.1. The distance relay is constantly looking in the line direction for the current and voltage measured at the relay location at the two ends of the line.where Z_F = fault impedance, V_F = fault voltage, and I_F = fault current.

$Z_p = Z_s$. (CT ratio/VT ratio) where Z_s = secondary impedance, Z_p = primary impedance.

The best way to study distance protection is to look for the simple balanced beam relay as shown in Fig. 18.6.2.

The voltage coil is supplied with the secondary voltage output of VT and the current coil is supplied with the current output from the CT. The torque resulting from the current coil is the operating torque but the torque resulting from the voltage coil is the restraining torque under normal operation of the power system. The two torques are balanced but during the fault condition

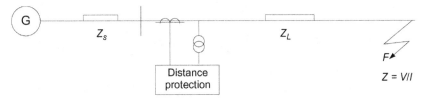

FIGURE 18.6.1   Distance protection principle.

$$Z_F = \frac{V_F}{I_F}$$

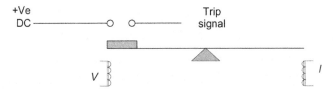

FIGURE 18.6.2   Simple balanced beam distance protection principle.

the voltage drops down and the current increases, which increases the operating torque, causing the relay to close the trip contact then transfer the positive DC to the trip relay, then later to line CB.

18.6.2   DISTANCE PROTECTION CIRCUIT ANALYSIS

Distance protection circuit analysis is illustrated in Fig. 18.6.3.

Impedance is measured by the relay at the relay location $Z_R = V_R/I_R = Z_L + Z$ load-at normal condition but in fault condition.

Impedance is measured by the relay at relay point: $Z_R = V_R/I_R = Z_F$.

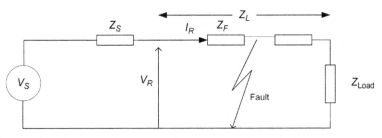

FIGURE 18.6.3   Circuit analysis of distance protection.

The relay will also operate if $Z_F < Z$, where $Z =$ the setting impedance set on the relay.

Increasing V_R has a restraining effect, so this voltage is called the restraining voltage and increasing I_R has an operating effect, and this current is called the operating current.

18.6.3 DISTANCE RELAY OPERATING CHARACTERISTIC

The relay operating boundaries are plotted, on an R/X diagram, its impedance characteristic is a circle with its center at the origin of the R-X coordinates and its radius will be the setting of the relay (relay reach) in ohms as shown in Fig. 18.6.4.

As shown in Fig. 18.6.5, the distance relay with plan characteristic is a nondirectional relay. This means the relay can trip for faults in a forward direction of the transmission line, and also for faults behind the relay in the reverse direction.

Another characteristic is used to give the relay the directional feature to see the faults on the forward direction of the line only. This is called Mho characteristic, as shown in Fig. 18.6.6.

The main advantages of this Mho characteristic are as follows:

1. It is a very popular characteristic;
2. It is very simple;
3. It has more immunity to power swings.

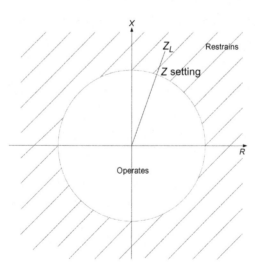

FIGURE 18.6.4 Distance relay simple plan characteristic.

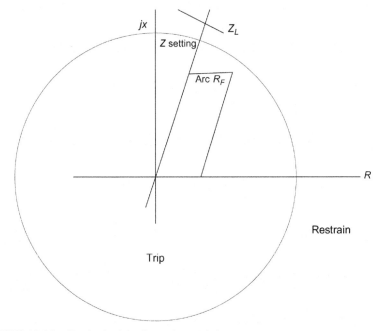

FIGURE 18.6.5 Simple plan impedance characteristic.

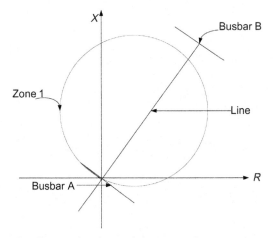

FIGURE 18.6.6 Mho distance relay characteristic.

However, the disadvantage of this characteristic is that it can work in a load condition, which is why we should take care of this point when setting the distance relay.

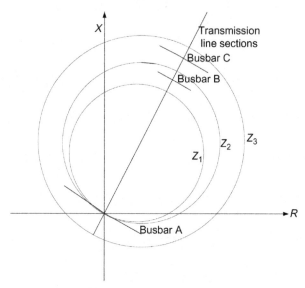

FIGURE 18.6.7   Distance relay protection zones.

To have good coordination in a distance relay application on high-voltage transmission lines we use three zones or more, as shown in Fig. 18.6.7.

Zone 1 is set to 80% of the line of Mho characteristic and is used to trip the line instantaneously in approximately one cycle based on a 50-Hz system (20 ms) this operating time is different from one manufacturer to the other and can be less than this depend on the relay speed of operation.

Zone 2 is set to about 120% of the line of Mho characteristic to cover the remaining part of the line (this means the remaining 20%) and a trip in a time delay set to about 500 ms.

Zone 3 is set about 225% of the line as offset-Mho characteristic to cover the other line sections but to trip in about 1000 ms this zone is also used as a starting element in the distance relay and this offset-Mho characteristic has the following advantages:

1. Working as backup protection for busbar faults in the reverse direction of the line;
2. Providing a local backup protection for close-up faults near the local end busbar of transmission line for faults which have a voltage near zero and the forward direction zone 1 cannot see;
3. Used in a permissive overreach blocking scheme in distance protection, which is explained in later sections of this chapter.

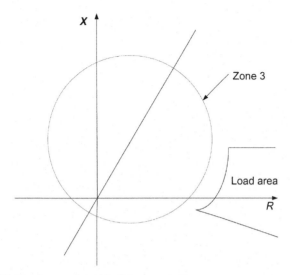

FIGURE 18.6.8   Load encroachment in distance protection.

When we set this zone we should take care that it does not encroach the load area, which causes unnecessary tripping, as shown in Fig. 18.6.8.

To overcome this problem we use a characteristic with a lenticular shape or eight and quadrilateral characteristics, which avoid the load area encroachment, as shown in Fig. 18.6.9.

As can be seen also from Fig. 18.6.5, the relay setting reach is set at about 80% only of the line length to prevent the relay from over-reaching or underreaching for the following reasons:

1. Due to the error in CT ratio and VT ratio, which can cause the relay to work in the wrong way in close-up faults near the remote end of the line.
2. Subtransient and transient short circuit (SC) currents have values more that the steady-state value of the SC current, which can cause the relay to under-reach. This means the measured impedance is less than the actual impedance of the fault, meaning that if the fault is outside the transmission line the relay will see it within the transmission line and trip in the wrong way and not follow the protection selectivity feature of the distance relay.
3. Due to high-resistance earth faults, which include a tower foot resistance in the path of the fault which can cause the relay to measure an impedance more than the actual impedance of the fault, this means for a boundary fault near the remote end of the transmission line if the fault is on the line the relay will see the fault as outside the

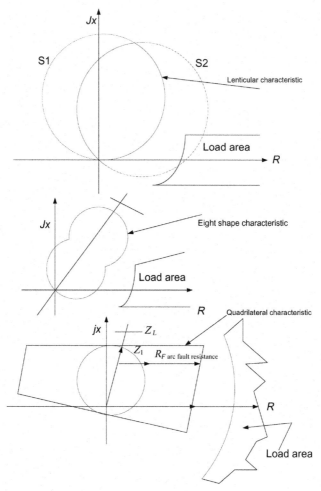

FIGURE 18.6.9 Special distance relay characteristic used as a remedial for load encroachments and arc fault resistance effect.

transmission line and will not trip for a fault which it is required to trip on (refer to Fig. 18.6.10).

As a fault on the transmission line with fault impedance Z with the effect of fault resistance, R_f will be seen by the relay as Z_x and cause the relay not to trip on this fault.

To avoid this problem of arc fault resistance we use the quadrilateral characteristic as shown in Fig. 18.6.9.

In general, distance relay has different characteristics depending on the application as shown in Fig. 18.6.11.

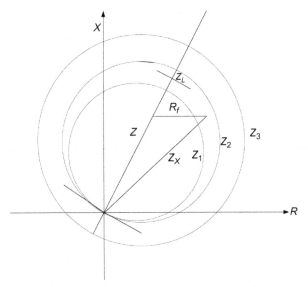

FIGURE 18.6.10   Over-reaching of distance protection due to the fault resistance effect.

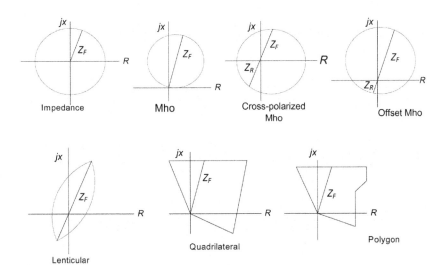

FIGURE 18.6.11   Different types of distance relay characteristic.

18.6.4   IMPORTANT CONSIDERATIONS IN DISTANCE PROTECTION

There are some important factors which should be considered when dealing with distance protection as described below.

18.6.4.1 Effect of System Impedance Ratio (SIR) on Distance Protection Measurements

The system impedance ratio (SIR) term refers to the ratio of source impedance of the power system behind the relay over the impedance of the transmission line, as follows:

$$\text{SIR} = \frac{Z_s}{Z_L}$$

As shown in Fig. 18.6.1, the transmission lines can be classified based on the value of SIR as follows:

1. Transmission line with high SIR ratio (SIR $>>>$);
2. Transmission line with low SIR (SIR $<<<$).

Refer to Fig. 18.6.12 for explanation of the effect of SIR on distance relay measurements at end A.

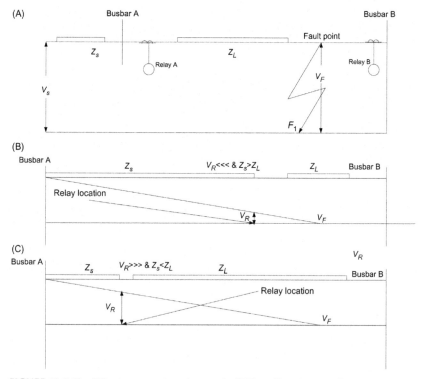

FIGURE 18.6.12 Effect or system impedance ratio (SIR) on distance protection measurements. (A) Impedance relay measurement; (B) short electrically line SIR more than 4; and (C) Long electrically line SIR less than 0.5.

For close-up faults near the remote end at busbar B, the relay can under-reach as the relay voltage V_R will be very small, and due to VT and CT errors, the relay will see the internal zone fault as a fault outside the line A—B. This will be for transmission lines with a high SIR ratio, as shown in Fig. 18.6.12B. There will be no problem for transmission lines with a low SIR ratio, as shown in Fig. 18.6.12C.

18.6.4.2 Distance Protection Setting for Parallel Lines

When we set a distance relay with two parallel transmission lines or a local infeed at the remote end, special consideration should be given, as explained here for local infeed at the remote end of the line, as shown in Fig. 18.6.13. where: Z_B = impedance from the end of the first line section A to the fault point F.

As shown in Fig. 18.6.13, we should consider the effect of the infeed current I_B on the distance relay setting as described below.

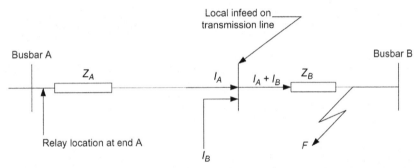

FIGURE 18.6.13   Effect of local infeed inside transmission line A—B on distance protection setting at end A.

The voltage measured by the relay at end A can be calculated as follows:

$$V_R = I_A\, Z_A + (I_A + I_B)\, Z_B$$

And the current measured by the relay at end A can be considered as follows:

$$I_R = I_A$$

Then, impedance seen by the relay will be as follows:

$$Z_R = (I_A \times Z_A)/I_A + ((I_A + I_B) \times Z_B)/I_A)$$
$$Z_R = Z_A + Z_B + I_B/I_A \times Z_B$$

As seen above, the relay with a setting of $Z_A + Z_B$ only will underreach.

The same consideration will be given in the case of a parallel transmission line distance relay setting, as shown in Fig. 18.6.14.

Where: Z_C = impedance from the end of the two parallel lines A and B to the fault point F.

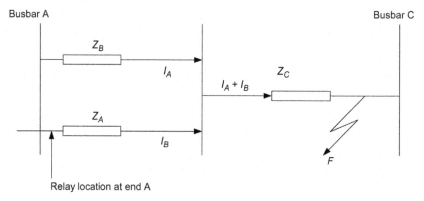

Relay location at end A

FIGURE 18.6.14 Effect of two parallel line sections (A and B) in transmission line A−C on a distance relay setting at end A.

As can be seen from Fig. 18.6.14, in the case of two parallel sections in a transmission line and one line connected to them as shown when we set the distance protection, the relay will be set as follows to avoid underreaching:

$$\text{The actual relay measurement is} Z_R = Z_A + \left(\frac{I_A}{I_A + I_B} \right) \cdot Z_C$$

As we set the relay to $Z_A + Z_C$ only the relay will underreach.

$$\text{Therefore,} Z_{R\ \text{SET}} = Z_A + \left(\frac{I_A}{I_A + I_B} \right) \cdot Z_C.$$

18.6.4.3 Weak Infeed in a Permissive Overreach Scheme

As explained in the permissive overreach transfer trip (POTT) scheme described in Section 18.6.5.3, in this scheme the POTT signal is sent by Zone 2, which is set to overreach the protected transmission line, but in the case of a weak infeed at one end of the transmission line, the relay R_1 at the weak infeed end will not see the close-up fault F_1 and will not trip and send the POTT signal to the strong end feed, as shown in Fig. 18.6.15.

To overcome this problem we set Z_3R_1 to cover the reach of Z_2R_2 in the reverse direction at end A, the weak infeed circuit will operate if these conditions occur:

1. Zone 3 of relay R1 (Z_3R_1) will not operate for fault F_1, this will confirm that there is no fault in the reverse direction at end A.
2. The voltage-measuring unit at relay A has lost voltage in one phase at least. Then the weak infeed circuit will operate and echo the POTT signal received from R_2Z_2 again to end B to accelerate R_2 to trip instantaneously and to trip R_1 instantaneously.

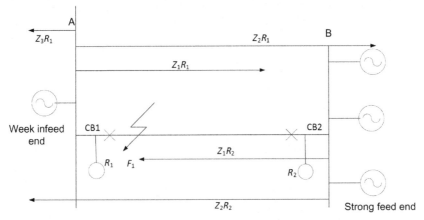

FIGURE 18.6.15 Effect of weak infeed in the permissive overreach transfer trip distance relay scheme.

18.6.5 DISTANCE PROTECTION TELECOMMUNICATION SCHEMES

As we can see previously, there is still 20% of the line section not protected instantaneously as we set Zone 1 to reach only up to 80% of the line for the reasons explained above. To overcome this problem we use telecommunication schemes which utilize a power line carrier, microwave radio signals, or the most recent being fiber-optic cables to transfer the protection signals through the transmission lines between both ends of the line to accelerate tripping in this remaining part of the line (Fig. 18.6.16).

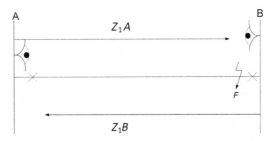

FIGURE 18.6.16 Direct transfer trip scheme.

18.6.5.1 Direct Transfer Trip Scheme (Underreach Scheme)

The fault F is seen by the relay at end B in Zone 1, then the relay at end B sends an acceleration direct trip signal to end A, at end A the fault F is seen in Zone 2 and Zone 3. If no received signal comes from end B then the relay will trip on Zone 2 with 500 ms, but if a received signal comes from end B then the relay will accelerate and trip instantaneously, as shown in Fig. 18.6.17, which shows the logic of trip for relay at end A.

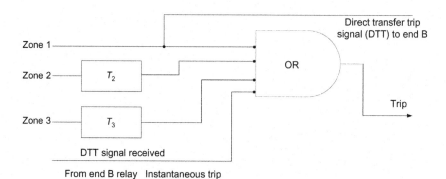

FIGURE 18.6.17 Trip logic for a direct transfer trip scheme.

The trip will occur instantaneously and includes only the time to receive the signal from the other end.

The disadvantage of this scheme is the possibility of false tripping by accidental or maloperation of signaling equipment.

18.6.5.2 Permissive Underreach Scheme

This more reliable and more secure system is a modification of the direct transfer trip scheme.

As shown in Fig. 18.6.18, a fault F is seen by relay B at end B of the transmission line, then a permissive trip signal is sent to end A, at end A if the relay A in Zone 2 sees the fault then the relay will accelerate Zone 2 and trip instantaneously, as shown in Fig. 18.6.19, which shows the trip logic of the permissive underreach transfer trip scheme.

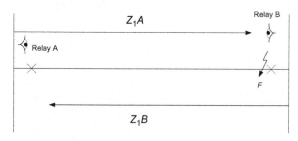

FIGURE 18.6.18 Permissive underreach transfer trip scheme.

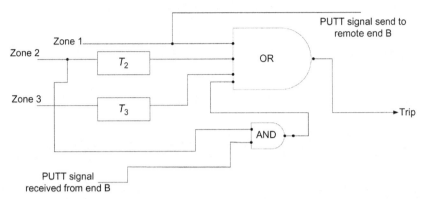

FIGURE 18.6.19   Trip logic of the permissive underreach transfer trip scheme.

18.6.5.3   Permissive Overreach Scheme (Directional Comparison Scheme)

In this scheme Zone 2 is set to 120% of the transmission line, as shown in Fig. 18.6.20.

- For a fault F, the relay at end B will see the fault and send the POTT signal to end A.

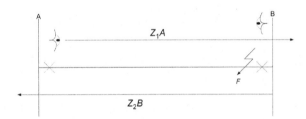

FIGURE 18.6.20   Permissive overreach transfer trip scheme.

- At end A, the relay Zone 2 will see the fault F and send the POTT signal to end B if Zone 2 of relay A sees the fault F and receives a POTT signal from remote end B then the relay A will trip instantaneously, if not the relay will trip in Zone 2 time (500 ms) (refer to Fig. 18.6.21 for POTT logic).

18.6.5.4   Zone Extension Scheme

In this scheme, the relay has two settings for Zone 1, The first is the basic setting at 80% of Z_L and the second is Zone 1 extension set to 120% Z_L, as shown in Fig. 18.6.22.

For a fault F, the normal setting of relay at end A is Z_1 extension, the relay A will trip on fault F. The autoreclose relay will reset the setting of the

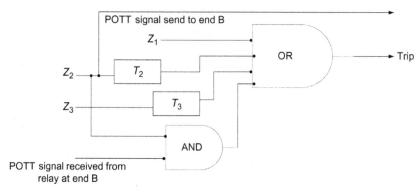

FIGURE 18.6.21 Trip logic for relay at end A.

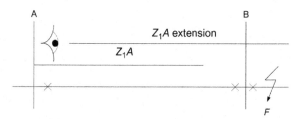

FIGURE 18.6.22 Zone extension scheme.

distance relay R_A to Z_1 basic = 80% Z_L and then will reclose the line AB successfully again to be in service.

This scheme is used with an A/R relay and can be implemented when there is a failure in teleprotection equipment in other teleprotection schemes as it is not dependent on a teleprotection signal from the remote end.

Refer to Fig. 18.6.23 for the scheme logic.

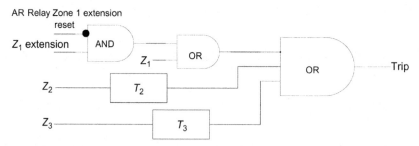

FIGURE 18.6.23 Zone extension scheme logic.

18.6.5.5 Acceleration Scheme

Previous designs of distance relays sometimes utilized one measuring element for Zone 1 and Zone 2, these were called switched distance relays, as illustrated in Fig. 18.6.24 for a fault, F.

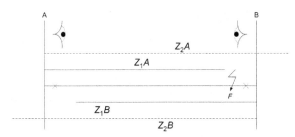

FIGURE 18.6.24 Acceleration scheme.

The relay at end B will see the fault and send a signal to relay at end A to switch the relay from Zone 1 reach to Zone 2 reach, the relay will trip instantaneously but it will trip on Zone 2 reach with 500 ms if no signal is received from remote end B—refer to scheme trip logic as shown in Fig. 18.6.25.

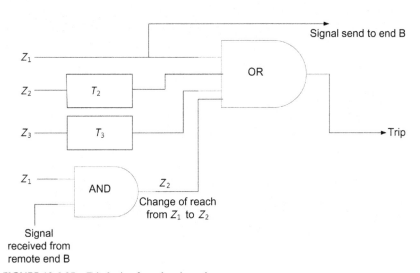

FIGURE 18.6.25 Trip logic of acceleration scheme.

18.6.5.6 Blocking Overreach Transfer Scheme

In this scheme, the teleprotection signal is used to block tripping of the relay at the remote end as shown in Fig. 18.6.26.

- In this scheme, Z_2 is set to 120% overreach of the transmission line AB, for fault F_1 at the remote end B, but outside line AB the fault will be seen by Zone 3 in the reverse of relay at end B then a blocking signal is sent to end A.

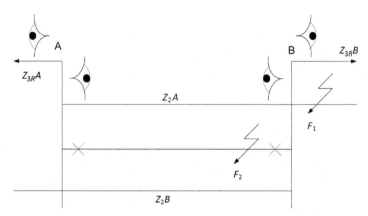

FIGURE 18.6.26 Blocking overreach transfer scheme.

- At end A, fault F_1 will be seen by Z_2A if a received blocking signal is received from remote end B. This means the relay will not trip on F_1, but if the fault is on the line side as F_2 no signal will be received from end B to block operation of relay at end A, then the relay at end A Zone 2 will trip instantaneously for fault F_2 (refer to Fig. 18.6.27 for the scheme trip logic).

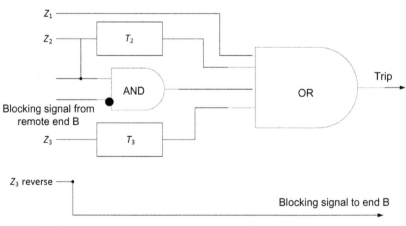

FIGURE 18.6.27 permissive overreach transfer trip blocking scheme.

18.6.6 DISTANCE RELAY INPUTS FOR CORRECT MEASUREMENT OF DIFFERENT TYPES OF FAULTS

The distance relay to measure the correct value of impedance should have the correct inputs as follows:

- Phase-to-phase faults;
  For correct measurement of Z_{L1} we should have the following inputs:
  Voltage input = phase to phase voltage;
  Current input = difference between currents in the faulted phases.

Fault	Voltage Applied	Current Applied
A-B	V_{AB}	I_A-I_B
B-C	V_{BC}	I_B-I_C
C-A	V_{CA}	I_C-I_A

- Phase-to-earth faults
  For correct measurement of Z_{L1} we should have the following inputs:
  Voltage input = phase to earth voltage on faulted phase;
  Current input = current in faulted phase + a proportion of the residual current depending on the ratio of zero to positive sequence impedance of the line.

Fault	Voltage Applied	Current Applied
A-E	V_{AE}	$I_A + K_N \times I_{RES}$
B-E	V_{BE}	$I_B + K_N \times I_{RES}$
C-E	V_{CE}	$I_C + K_N \times I_{RES}$

Where K_N = residual earth fault compensation factor = $\frac{(Z_{L0} - Z_{L1})}{3Z_{L1}}$, I_{RES} = residual current = $(I_A + I_B + I_C)$.

In general practice as a guide K_N for overhead lines = 0.6 and for cables $K_N = 0.3$.

18.6.6.1 Neutral Impedance Compensation

For a single phase to ground fault the total earth loop impedance is given by:
$$Z_T = \frac{(Z_1 + Z_2 + Z_0)}{3}$$

$$Z_T = \frac{(Z_1 + Z_2 + Z_0)}{3} = Z_1 + Z_N$$

$$Z_N = \frac{(Z_1 + Z_2 + Z_0)}{3} - Z_1$$

$$Z_N = \frac{(2Z_1 + Z_0)}{3} - Z_1$$

$$Z_N = \frac{-Z_1}{3} + \frac{Z_0}{3}$$

$$Z_N = K_N Z_1$$

where

$$K_N = \frac{(Z_0 - Z_1)}{3Z_1}$$

18.6.7 POWER SWING BLOCKING RELAY

This feature along with distance protection to block the distance function in the relay during power swing conditions. It is utilizes an outer characteristic for relay Zone 3 and the rate of change of impedance between the two characteristics means z power swing and Zone 3 will be monitored if the change of impedance is slow and the impedance lies between the two characteristics for a time longer than the power swing time then the power swing relay will block the distance function of the relay, but if the rate of change of impedance is very fast and the impedance passes between the two characteristics very quickly then the power swing relay recognizes this condition as a fault and does not block the distance function and allows the relay to trip (refer to Fig. 18.6.28).

The basic criterion for power swing condition is $\frac{dZ}{dt}$, if it is high then it is a fault condition but if it is low then it is a power swing condition.

18.6.8 VOLTAGE TRANSFORMER SUPERVISION (FUSE FAILURE)

This feature is incorporated with the distance relay to monitor the secondary voltage of the VT and to give an alarm in case of losing voltage or to block the operation of the distance relay.

This function can work in a zero-sequence principle or negative-sequence principle to monitor the secondary voltage of the VT and in case of loss of voltage to block operation of different relay functions which depend on voltage as distance protection, synchro check function, and under voltage function, etc., the blocking function is issued after a time delay of the order of 3−5 seconds.

The criteria for fuse failure detection are as follows:

1. Detection based on the rise of the value of zero-sequence voltage $3V_0$ or negative-sequence voltage V_2 above the setting value.
2. The value of the zero-sequence currents $3I_0$ or negative-sequence current is less than the setting value.

However, in case of detecting high unsymmetrical zero-sequence currents $3I_0$ during the blocking operation of distance function due to fuse failure detection then this block will be released.

Also, some manufacturers of relays are switching the distance function during a fuse failure blocking condition to an emergency overcurrent relay.

(A)

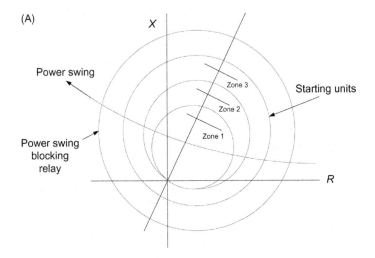

(B)

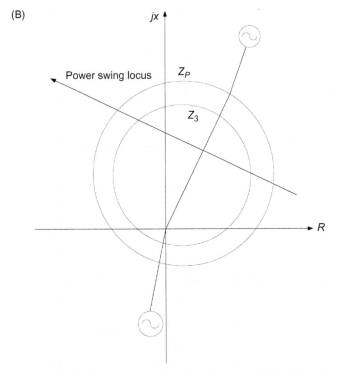

FIGURE 18.6.28 (A) Distance relay characteristic and power swing blocking relay characteristic. (B) Power swing blocking relay characteristic.

18.6.9 SWITCH ON TO FAULT FEATURE IN DISTANCE RELAY

This feature in distance protection is activated after manual closure of a CB and is designed to bypass the distance relay measuring element and switched to a high-speed overcurrent function in case there is a close-up three-phase fault near the relay in the local busbar of the relay, as in practice maintenance people forgot the local three-phase earthing on the line under maintenance inside the substation, as in this case the relay voltage will be near zero and the fault is a three-phase close-up fault.

The criteria for detecting the switch to fault (SOTF) condition are by voltage-level detector and current-level detector, for example, setting these two detectors is as follows:

$V_{ph} < 75\%$ V_n and $I_{ph} > 5\%$ In, as V_n is the nominal voltage and I_n is the nominal current of the system the SOTF will be energized for about 500 ms and then reset after this time. If the conditions above are detected during manual closure of the breaker during this time the SOTF will trip the line instantaneously.

18.6.10 STUB PROTECTION FUNCTION

There are short sections of the current path within a substation that are not properly protected by the main protection system. These sections are called stubs or blind spots protection or minimum zone protection, and they are usually between the CB and the CT. The main protection system measures the current of the CT and, if a fault is detected, a command is generated to open the CB. If, however, the fault is between the CB and the CT, then opening the CB cannot clear the fault; it is fed via the CT from the other side of the protected object. This location is within the back-up zone of the other side protection and, accordingly, it is cleared by a considerable time delay. The task of the stub protection function is to detect the fault current in the open state of the CB and to generate a quick trip command to the other side CB. Another usual application is in the one-and-a-half CB arrangement. Here the CTs are located either before or after the CBs. Additionally, the VT is either on the bus side or on the line side of the isolator. In the last case, the stub is also the section between the CBs and the open line isolator, since if a fault occurs in this section, the detected voltage is independent of the fault; it is unchanged and cannot be applied for the distance protection. The stub protection function is basically a high-speed overcurrent protection function that is enabled by the open state of a CB or may be an isolator. If any of the phase currents is above the start current and the binary status signal activates the operation, then after a user-defined time delay the function generates a trip command. The function can be disabled by programming the blocking signal in new

IEDs. The stub protection in a breaker-and-a-half application. With the breaker-and-a-half configuration an operating mode is possible whereby the feeder is out of service while both CBs in the diameter remain closed. This is the case when DS1 is open, while CB1 and CB2 are closed in the SLD is shown in Fig. 18.6.29.

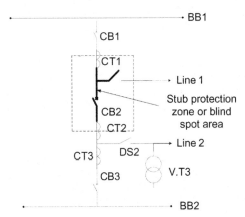

FIGURE 18.6.29   Stub protection in a one-and-a-half circuit breaker busbar arrangement.

All faults in the zone between CB1, CB2, and DS1 are stub faults. The stub protection boundary is defined by the location of the CTs, as shown above.

18.6.11   AUTORECLOSING FUNCTION IN DISTANCE RELAY

This relay is used in a distance protection scheme to reclose the lines automatically after transient faults trip by distance relay.

18.6.11.1   Autoreclosing Important Definitions

Eighty percent up to ninety percent of transmission line faults are transient faults. This provides a need to use an autoreclosing system. There are basic terms and definitions in autoreclosing systems as follows:

Arcing time:

The time between the instant of separation of the CB contacts and the instant on extinction of the fault arc.

Closing impulse time:

The time during which the closing contacts of autoreclosing relay are made.

Closing time:

The time from the energizing of the CB closing circuit to the making of the CB contacts.

Dead time of autoreclosing relay:

The time between the autoreclosing scheme being energized and the operation of the contacts which energize the circuit-breaking closing circuit. On all but instantaneous or very high-speed reclosing schemes, this time is virtually the same as the CB dead time.

Dead time of CB:

The time between the fault arc being extinguished and the CB contacts remaking.

High-speed reclosing scheme:

A scheme whereby a CB is automatically reclosed within 1 second after a fault trip operation.

Lock-out of autoreclosing relay:

A feature of the autoreclosing scheme which, after tripping of the CB, prevents further automatic reclosing.

Delay autoreclosing (DAR) scheme:

A scheme whereby the automatic reclosing of the CB following a fault trip operation is delayed for a time in excess of 1 second.

Multishot reclosing:

An operation sequence providing more than one reclosing operation on a given fault before lock-out of the CB occurs.

Opening time of a CB:

The time from the energizing of the trip coil until the fault arc is extinguished.

Operation time of protection:

The time from the inception of the fault to the closing of the tripping contacts of the relay. Where a separate tripping relay is employed, its opening time is included.

Reclaim time:

The time following a successful closing operation, measured from the instant the autoreclose relay makes contact, which must elapse before the autoreclose relay will initiate a reclosing sequence in the event of a further fault incident.

Single shot autoreclosing:

An operation sequence providing only one reclosing operation, lock-out of the CB occurring on a subsequent tripping.

Refer to Figs. 18.6.30 and 18.6.31 for an autoreclosing cycle for transient and permanent faults.

18.6.11.2 Autoreclosing for Transmission Lines

For 500-kV lines, only lines to ground faults are candidates to allow high-speed autoreclose.

High-speed auto reclosing (unsupervised) should be applied where it supports system stability or if load continuity is a concern.

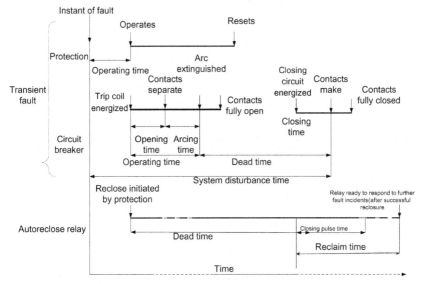

FIGURE 18.6.30 Autoreclosing relay cycle for transient faults.

Reclosing on a noncritical transmission line should be performed at one end first by the live bus/dead line. The other end(s) should be reclosed through synchronizing equipment.

Reclosing on transmission lines near generation must consider the impact on the machine shaft torque as well as system stability. If possible, utilize 10 seconds or longer delayed reclosing at the remote terminal to limit the interaction with shaft oscillations and only reclose through system synchronization at the generation terminal.

Where system stability is a concern, the line may be tripped single phase and high-speed autoreclosed. (Unbalance effects on zero-sequence protection may require some evaluation by the applicant.) All multiphase faults should trip three phase and consideration should be given to blocking autoreclosing.

High-speed autoreclose should only be applied for high-speed trips in communications-aided Zone 2 trips in distance protection for stability-critical and load-sensitive lines. High-speed autoreclosing should be disabled or blocked for loss of communication channels. Lines that do not use communications schemes should be reclosed through synchronizing or voltage supervision. (The speed of clearing has a significant effect on the rotor angle of the generators.) The faster that the fault is cleared, the more time that is available to provide more retarding torque to bring the rotor angle back into synchronism when the line is reclosed.

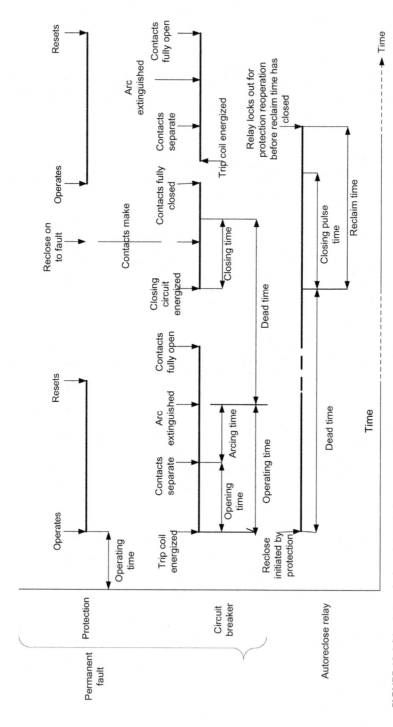

FIGURE 18.6.31 Autoreclosing relay cycle for permanent faults.

18.6.11.3   Autoreclosing and Other Equipment

Different types of autoreclosing equipment have been developed to meet the varying requirements on the dead time and number of reclosing shots:

1. One high-speed reclosing shot;
2. One high-speed reclosing shot followed by one or more delayed reclosing shots;
3. One delayed reclosing shot;
4. Several delayed reclosing shots.

The equipment is supplied as separate units or incorporated in the line protections. It can be easily supplemented and adapted for different breaker schemes, for example, the one-and-a-half breaker scheme and the double breaker scheme.

In addition, other, more sophisticated types of equipment have been developed for automatic system restoration after a disturbance. Such equipment may sometimes work together with synchro-check and synchronizing equipment to restore the network.

Automatic equipment has also been developed for automatic load restoration after load shedding.

18.6.11.4   Distribution Networks with Radial Feeders

Problems of system stability or synchronizing requirements do not occur with radial feeders with load only and therefore they can easily be provided with three-phase autoreclosing equipment. This gives certain benefits such as short outage times and the possibility to save on personnel. If instantaneous tripping and high-speed autoreclosing are utilized, damage to the line in connection with a transient fault can be limited. When feeders consist partly of underground cables, the suitability of autoreclosing should be considered, since the faults occurring in a cable are generally permanent. If there is only a small risk of the cable being damaged by excavators, autoreclosing may nevertheless be justified, since cable faults do not arise as frequently as faults on overhead lines.

18.6.11.5   Autoreclosing in Strong High-Voltage Networks

If the network is sufficiently strong, three-phase high-speed autoreclosing can be utilized without the synchronism or stability being disturbed. Simple equipment without facilities for checking the voltage, phase angle, and frequency difference can then be used. Protective relays, breakers, etc., will be simpler than for single-phase autoreclosing and no zero-sequence and negative-sequence currents are obtained during the dead time. When only three-phase delayed autoreclosing is utilized, no problems are experienced with the deionizing time and breaker operating time. If oscillations occur in the network, they generally have time to become stabilized before a reclosure. There are somewhat greater chances of a successful reclosure with three-phase DAR than with only high-speed autoreclosing. When a line has been tripped, the voltage in the network increases or decreases owing to the change in the load. Transformer tap changers may then start to operate with the risk of being damaged, if reclosing on to a short circuit occurs during the switching process. The delay in the voltage regulation equipment should therefore be checked. This delay should be longer than the system disturbance time. Several reclosing shots may be justified in the case of, for example, radial lines. However, the advantages of the increased probability of a successful reclosure must be weighed against the disadvantages in the form of greater damage at the site of the fault and higher breaker maintenance costs.

Subchapter 18.7

Generator Protections

18.7.1 DIFFERENTIAL PROTECTION

This main protection of stator winding compares current values at the end of each phase winding and operates when a predetermined percentage difference is detected. This protects the winding against earth and phase to phase faults. This protection can be of a high-impedance type or low-impedance biased type, as shown below in Fig. 18.7.1.

There is a scheme called overall differential protection, which can protect the generator unit, main transformer, and also the unit transformer, as shown in Fig. 18.7.2.

18.7.2 STATOR EARTH FAULT PROTECTION (90%)

This system protects 90% of generator winding, as shown in Fig. 18.7.3.

For unearthed generators:

51 N time delayed;
50 N instantaneous only for large units;

Relays must not operate for residual current (I_{Resd}) caused by CT output summation error and internal capacitive current of external earth faults.

For earthed generators:

51 N must be graded with other earth fault relays on the power system;
51 N time delayed set to 5% of generator resistor rating;
50 N instantaneous only for large units set to 10% of generator resistor rating.

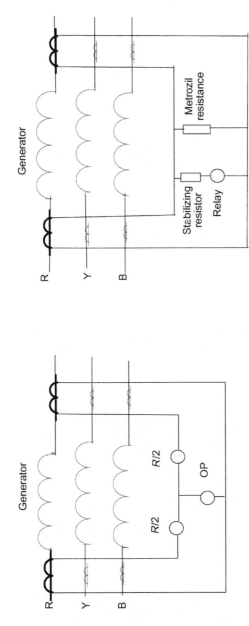

(<Z) High impedance differential relay for R phase

(>Z) Low impedance bised differential relay for R phase

R = restraining coil

OP = operating coil

FIGURE 18.7.1 Generator differential relays.

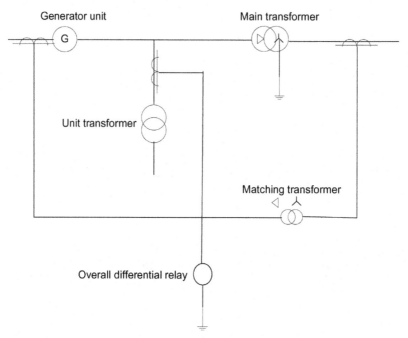

FIGURE 18.7.2   Overall differential scheme for generator units.

18.7.3   STATOR EARTH FAULT PROTECTION (100%)

This scheme is applied for large machines only.

A low-frequency injection method is shown in Fig. 18.7.4:

Complete protection for stator windings;
High cost of injection equipment is a disadvantage.

Third harmonic neutral voltage scheme as shown in Fig. 18.7.5:

27 Third harmonic undervoltage relay;
59P terminal voltage check permits for trip if circuit breaker is open but a terminal voltage is present;
Neutral overvoltage protection operates at a fundamental frequency voltage of 50 Hz.

The trip logic is shown in Fig. 18.7.6.

As shown in Fig. 18.7.7, the overlapping of the operation of fundamental frequency overvoltage relay-59 and the third harmonic undervoltage relay-27 will cover by 100% the protection for the stator.

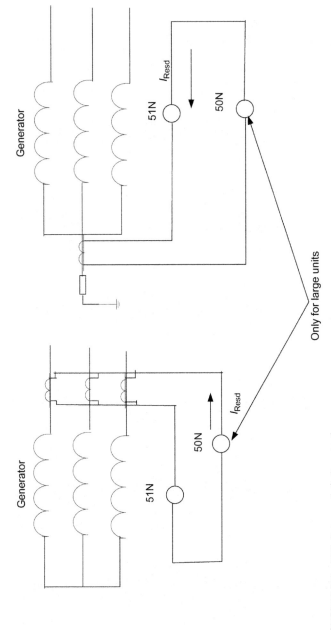

FIGURE 18.7.3 90% Stator earth fault protection.

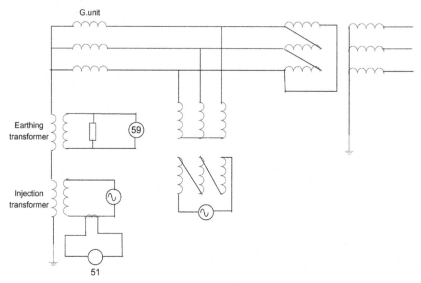

FIGURE 18.7.4   Low-frequency injection method for 100% stator earth fault protection for generators.

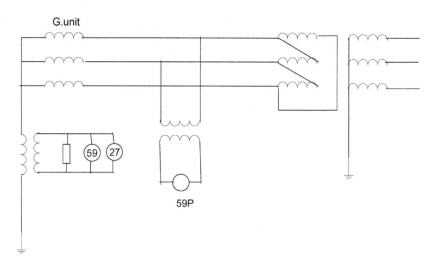

FIGURE 18.7.5   100% Stator EF scheme with third harmonic voltage scheme.

18.7.4   RESTRICTED EARTH FAULT PROTECTION

This is a high-impedance protection, and it is instantaneous and protects 90%−95% of generator winding; all CTs are similar and should be class *X*.

Refer to Fig. 18.7.8 for Restricted Earth Fault (REF) protection for a generator.

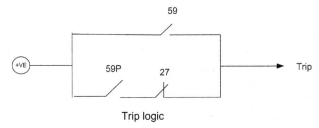

Trip logic

FIGURE 18.7.6 Trip logic of the relay.

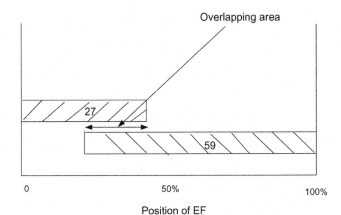

Position of EF

FIGURE 18.7.7 Overlapping of 59 and 27 relays for 100% stator EF protection for generators.

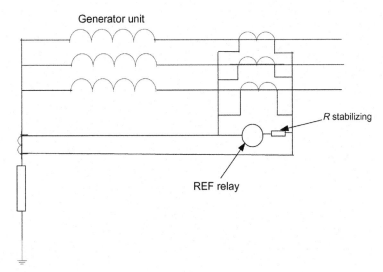

FIGURE 18.7.8 REF protection for generators.

18.7.5 OVERCURRENT PROTECTION

This protection is the only protection for small units. It provides protection for the generator and power system, therefore a time delay is required for coordination.

Refer to Fig. 18.7.9 for the OC protection.

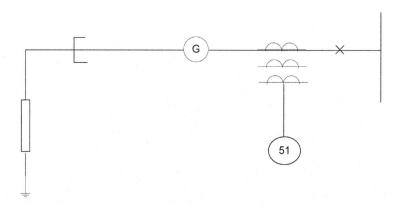

FIGURE 18.7.9 OC protection for generators.

The variable response of a generator for change in excitation and voltage regulation led to use of overcurrent relays with different characteristics based on the generator terminal voltage value.

There are two types of OC protection applied to generators:

1. Voltage-controlled overcurrent protection

During normal overloads, system voltage is near normal and the relay operates on long inverse time characteristics but under close-up fault conditions the busbar voltage falls and the voltage relay will change the relay characteristics to standard IDMT characteristics. Refer to Fig. 18.7.10 for voltage control overcurrent relay.

2. Voltage-restrained overcurrent protection

In this type of relay, the relay characteristic is modified continuously based on the generator terminal voltage. Refer to Fig. 18.7.11 for voltage-restrained overcurrent protection.

18.7.6 BACKUP DISTANCE PROTECTION

This impedance relay is used to provide backup protection for faults on a power system which is not cleared by the power system protection. This protection operates to trip generators after a time delay, as shown in Fig. 18.7.12.

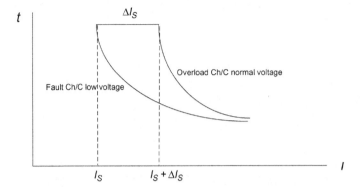

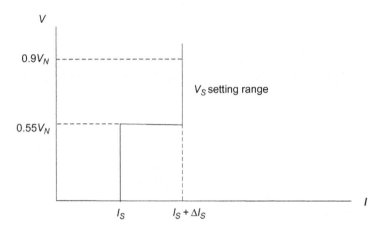

When voltage change the relay Ch/C changes and I_{Set} become $I_{Set} + \Delta I_S$

FIGURE 18.7.10   Voltage-controlled OC relay for generators.

18.7.7   REVERSE POWER PROTECTION

Reverse power protection is taken against generator prime mover failure. The motoring allowable for generator units as follows:

Steam turbine and gas turbine units: 5% only;

Diesel-driven sets: 25% only;

Hydro units reverse power protection is not required.

In this case there is a field current but there is no steam to drive the turbine, this condition leads to reversing the active and reactive power and the generator runs as a synchronous motor, leading to a condition in which the turbine blades are not cooled and can be distorted and damaged.

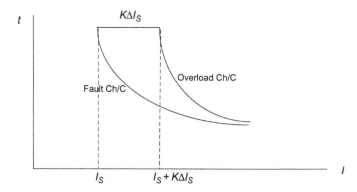

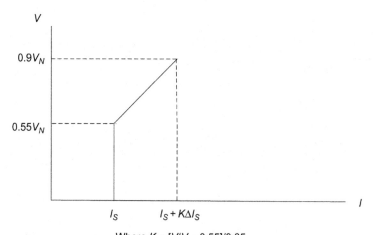

Where $K = [V/V_N - 0.55]/0.35$

Relay Ch/C continuously changed based on voltage of generator

FIGURE 18.7.11 Voltage-restrained OC relay for generators.

18.7.8 LOSS OF EXCITATION (FIELD FAILURE) PROTECTION

In this case there is steam to drive the turbine but no field current, and then the machine absorbs reactive power Q from the system, leading to loss of synchronism for the generator. To detect this condition an offset-Mho distance relay as shown in Fig. 18.7.13 was used.

18.7.9 OUT-OF-STEP (POLE SLIPPING) PROTECTION

Two Ohm relays are used to detect the out-of-step condition of the generator (refer to Fig. 18.7.14).

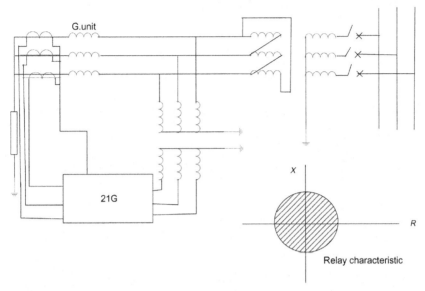

FIGURE 18.7.12 Backup distance relay for generators.

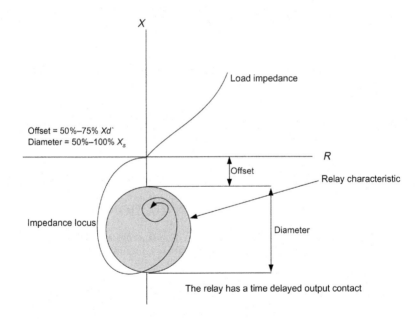

Offset Mho distance relay characteristic to monitor the impedance at geneartor terminals

FIGURE 18.7.13 Loss of excitation distance relay protection for generators.

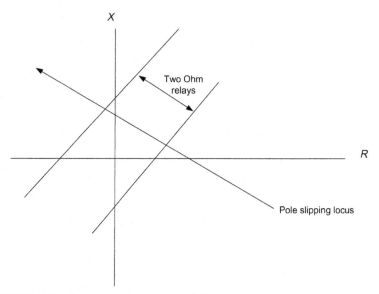

FIGURE 18.7.14   Out-of-step relay characteristic for generators.

18.7.10   NEGATIVE-PHASE SEQUENCE CURRENT I_2 PROTECTION

Unbalanced loading of generators which generates negative-sequence currents in the stator which cause a controlled-rotating magnetic field which cuts the rotor at twice the synchronous speed inducing double-frequency current in the field system and rotor body, which leads to eddy currents causing overheating, there are two ratings for I_2 for machines as follows:

I_2R = continuous Negative Phase Sequence (NPS) rating;
I_2^2t = short time NPS rating.

18.7.11   ROTOR EARTH FAULT PROTECTION

A single fault will not cause any problem but a second fault in the field winding of the rotor could cause serious damage including:

Heating-conductor burning;
Flux distortion in the field leading to generator vibrations.

A DC injection method is used by injecting a DC voltage into field windings to detect the earth faults, as shown in Fig. 18.7.15.

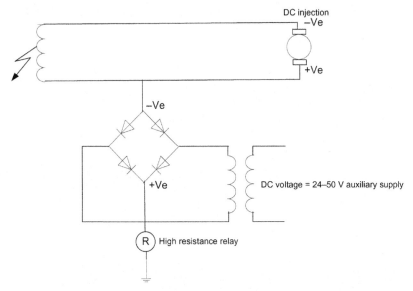

FIGURE 18.7.15   Rotor earth fault protection for generators.

18.7.12   FREQUENCY PROTECTION

Over-frequency protection is:

Caused by a sudden loss of load;
Can be considered as a backup protection for governor failure.

Underfrequency protection is caused by excessive overload and can cause a rise in:

Overfluxing of the stator core;
Plant drives operate at lower speeds affecting generator output;
Mechanical resonant condition in turbines.

This protection initiates load shedding.

18.7.13   VOLTAGE PROTECTION

Overvoltage protection is a condition of overvoltage which occurs when the prime mover overspeeds due to a sudden loss of load, and the voltage regulator is defective.

This condition can cause failure of insulation and also causes overexcitation.

With undervoltage protection there is no danger for the generator but it causes stalling of motors in the generator's service motors.

18.7.14 INTERTURN FAULT PROTECTION

Normal differential protection cannot detect this fault which is rare in generators and quickly develops to an earth fault.

The generator can be protected against this fault by zero-sequence voltage protection, as shown in Fig. 18.7.16.

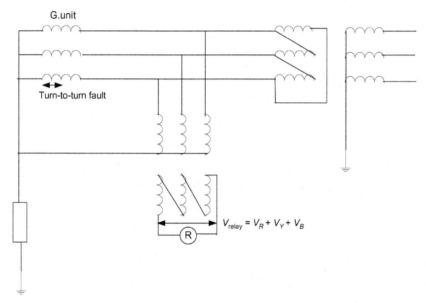

FIGURE 18.7.16 Interturn fault protection for generators by zero-sequence voltage.

Also, this fault can be detected by differential protection in double-wound generators, as shown in Fig. 18.7.17.

18.7.15 POWER FACTOR RELAY (32*R*) PROTECTION

This protection works for reverse reactive power (Q) as is the case in loss-of-field current for generators (the generator absorbs Q (MVAR) from the power system).

18.7.16 CONCLUSION

Fig. 18.7.18 depicts a typical generator protection.

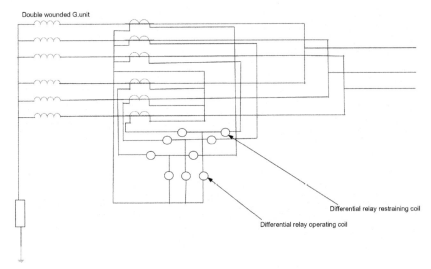

FIGURE 18.7.17 Differential protection in double-wound generators for interturn faults.

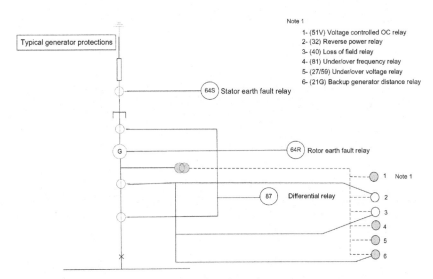

FIGURE 18.7.18 Typical generator protective relays scheme.

Subchapter 18.8

Motor Protection

18.8.1 INTRODUCTION

18.8.2 Different Types of Motor Protections

1. Instantaneous earth fault protection

   Motors with powers greater than 50 HP, supplied from an earthed system, should be protected against earth faults to reduce fault damage and accident risk, especially motors not protected by differential protection.

   The rotor is also protected against earth faults.

2. Differential protections

   These are normally fitted in machines of 1000 HP and above.

   Transverse differential protection can be used against interturn faults, where the stator windings are divided into two or more circuits.

3. Overloading and stalling protection

   Thermal relays are employed for overloading, also a separate stalling relay is employed for motor stalling conditions.

4. Instantaneous high-set overcurrent protection

   This can be included with thermal overload relays.

5. Unbalance protection

   Unbalanced or negative-phase sequence protection should be used as rotor heating due to unbalanced currents being a function of the negative-sequence component of line currents.

   When a motor is stalled due to loss of one phase, the heating is concentrated in one part of the rotor and the instantaneous negative-sequence unit can give complete protection.

6. Protection against restoration of supply

   Synchronous machines must be protected against this condition because it can go out of step with supply after an interruption. A sensitive underfrequency relay is used for this condition.

   Induction motors are protected against this condition by no-voltage release on the starter because the motor terminal voltage falls rapidly on loss of supply.

7. Reverse-phase sequence protection

   A reverse-phase sequence and undervoltage relay can be used to detect this condition.

8. Bearing failure protection

   Failure of a bearing can cause motor stalling. A defective bearing is indicated by temperature rise and vibration and a small rise in motor

current. A temperature-sensing device embedded in bearings gives adequate warning.
9. Loss of synchronism and field failure in synchronous motor protection

Synchronous machines subjected to sudden mechanical overloads should be protected against loss of synchronism, which can also be caused by a reduction of field current or supply voltage.

Where the overloads period is short a loss of synchronism relay can be arranged to disconnect the field supply only, so that the machine runs as an induction motor during overload then the relay resynchronizes the machine automatically after overload periods.

Field failure relay should be provided for synchronous machines larger than about 500 HP which are not provided with loss of synchronism protection, preventing overheating of the rotor after continuous operation as an induction motor. An undercurrent relay in the field circuit gives satisfactory protection.

Damper winding thermal protection may be used as an alternative to field failure protection if the machine is required to run for long periods as an induction motor.

18.9 Switchgear (Busbar) Protection

18.9.1 IMPORTANCE OF BUSBARS

Busbars are the most important component in a distribution network. They can be open busbars in an outdoor switch yard, up to several hundred volts, or inside a metal-clad cubicle restricted within a limited enclosure with minimum phase-to-phase and phase-to earth clearances. We come across busbars which are insulated, as well as those which are open and are normally in small-length sections interconnected by hardware.

They form an electrical "node" where many circuits come together, feeding in and sending out power (see Fig. 18.9.1).

From Fig. 18.9.1, it is very clear that for any reason the busbars fails, it could lead to a shutdown of all distribution loads connected through them, even if the power generation is normal and the feeders are normal.

18.9.2 BUSBAR PROTECTION

Busbars are frequently left without protection because it is very rare to have faults, especially metal-clad switchgear, and it is protected by backup protection, it can be protected by a separate busbar protection but it is very expensive due to the cost of CTs and relays. Also, any false operation will cause the complete system to trip.

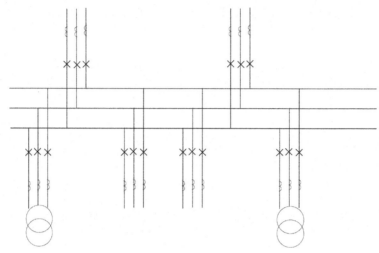

FIGURE 18.9.1 Schematic illustrating area of a busbar zone.

Good busbar protection will have the following features:

Main and check zones;
Block in case of CT open circuited;
Very fast;
Very good stability;
Selective tripping for the faulted section only;
Good isolation facilities and short circuit facilities for CTs during testing.

18.9.3 BUSBAR PROTECTION TYPES

1. Frame leakage;
2. Busbar blocking;
3. Busbar differential protections
   A. Low-impedance biased differential
   B. High-impedance differential;

18.9.3.1 Frame Leakage Busbar Protection

This involves measurement of the fault current from the switchgear frame to the earth. It consists of a CT connected between frames to earth points and energizes an instantaneous earth fault relay to trip the switchgear. It generally trips all the breakers connected to the busbars.

Care must be taken to insulate all the metal parts of the switchgear from the earth to avoid spurious currents being circulated. A nominal insulation of 10 MΩ to earth is sufficient. The recommended minimum setting for this protection is about 30% of the minimum earth fault current of the system (see Fig. 18.9.2).

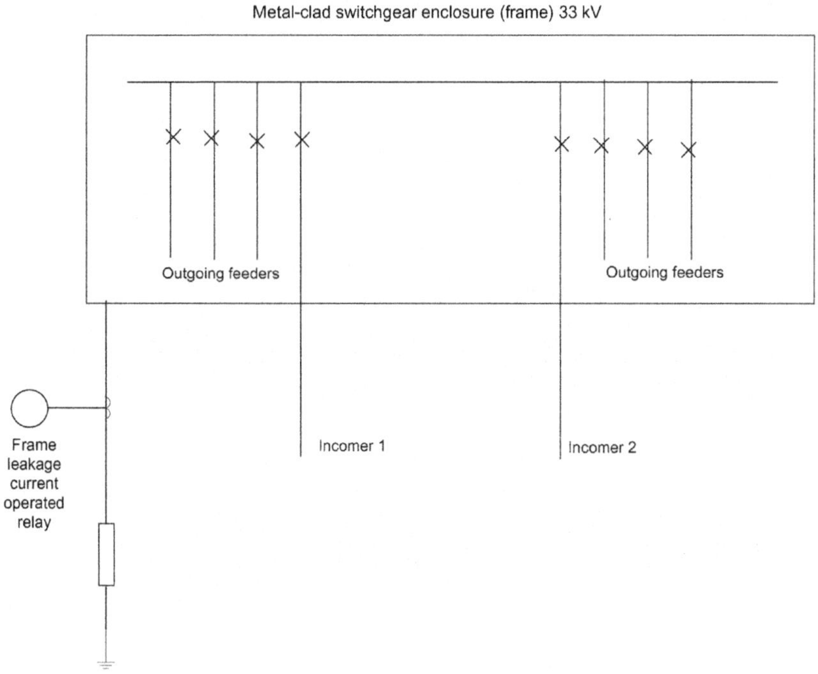

Metal-clad switchgear enclosure (frame) 33 kV

FIGURE 18.9.2   Schematic connections for frame leakage protection.

18.9.3.2   Busbar Blocking System

A three-busbar blocking system is applied for simple busbars with incomer and outgoings, as shown in Fig. 18.9.3. If the fault is in the outgoing feeders, as Fault number 2 (F2), this will block the operation of the busbar relay but if the fault in the busbars is as Fault number 1 (F1) this will issue a trip signal to the incomer feeder as there are no blocking signals from outgoing feeders being issued.

The setting of the incomer overcurrent relay is higher than the setting of the outgoing feeders.

18.9.3.3   BusBar Differential Protection

Three-busbar differential protection is used to protect the busbars sections separately as different zones, as shown in Fig. 18.9.4.

As shown in Fig. 18.9.4 each section of the busbars is protected by a separate unit and the trip is confirmed by a check zone which protects the complete busbar (refer to the above trip logic in Fig. 18.9.4).

This type of busbar protection has two types: low-impedance biased differential protection and high-impedance differential protection, as described here.

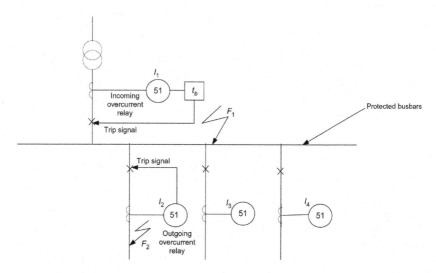

I_1 Incoming overcurrent relay has a setting more than the outgoing feeder overcurrent relays

I_2, I_3, I_4 Outgoing overcurrent relay has a setting Less than the incoming feeder overcurrent relay

T_b = blocking timer

FIGURE 18.9.3   Busbar blocking scheme.

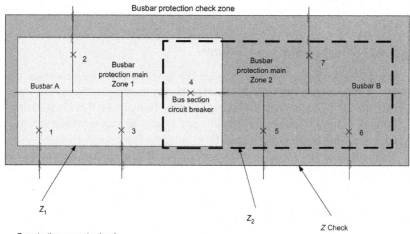

Z_1 protection covers busbar A
Z_2 protection covers busbar B
Z Check protection covers busbar A and busbar B

Busbar protection trip logic

FIGURE 18.9.4   Zoned busbar (switchgear) protection.

18.9.3.3.1   Low-Impedance Busbar Prottection

Low-impedance busbar protection uses the Merz-price circulating current principle for biased differential protection to detect a fault in the busbar zone, as shown in Fig. 18.9.5.

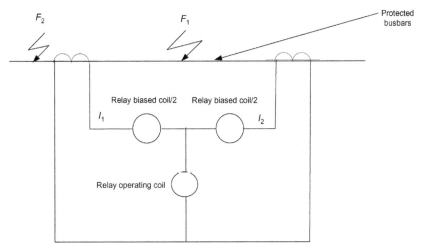

FIGURE 18.9.5   Low-impedance biased differential protection for busbars.

As shown in Fig. 18.9.5, for a fault F_1 the relay will trip, but for F_2 the relay will be stable due to the effect of the relay restraining coil. The operating time of this relay is in the range of 20 ms.

18.9.3.3.2   High-Impedance Circulating Current Protection

Refer to Figs. 18.9.6 and 18.9.7 for a high-impedance differential protection scheme.

V relay $= I_1 \times (R_{CT1} + R_{L1})$, V relay $= I_R \times R$

$R = V$ relay$/I_R$ then $R = (I_1 \times (R_{CT1} + R_{L1}))/I_R$

$$R = \frac{I_1 \times (R_{CT1} + R_{L1})}{I_R}, R_{st} = R - R \text{ relay}$$

where V relay $=$ voltage across the relay during the external fault F; $R_{st} =$ stabilizing resistor; $R_{CT} =$ CT winding resistance; $R_L =$ lead resistance.

For maximum external fault **F**, the current in relay R is theoretically zero, where R_{L1}, R_{L2} are lead resistances, R_{CT} is the CT resistance, if one CT becomes fully saturated in one side its secondary EMF will become zero and this can be represented as short circuited across its magnetizing impedance. This is the worst case for stability of high-impedance relay and the relay must be stable under this condition for the maximum external fault. Under this condition the current I_1 will pass through the relay circuit and the saturated CT−SC branch.

Where $I_R =$ relay current setting, $R =$ relay circuit impedance, $V_s =$ setting voltage, $V_R =$ relay voltage,

Then $V_R = I_1 (R_{CT} + R_{L1})$

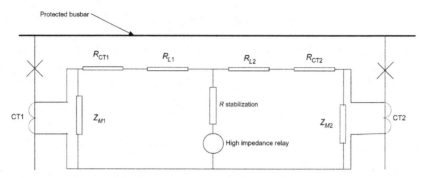

FIGURE 18.9.6 Simple high-impedance circulating current scheme with two current transformers.

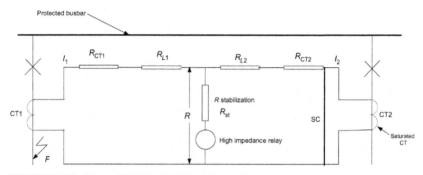

FIGURE 18.9.7 External fault in a high-impedance scheme.

Adjusting the relay impedance so that the voltage required to operate the relay is greater than the voltage V_R:

$$V_s > V_R$$
$$I_R R > I_1 (R_{CT} + R_{L1}) \text{ then } R > I_1/I_R (R_{CT} + R_{L1})$$

In order to obtain the required value of R it is usually necessary to use an additional resistor called a stabilizing resistor (R_{ST}) in series with the relay coil (R) relay, so the required stabilizing resistor is as follows:

$$R_{ST} = R - R \text{ relay}$$

In case of an internal fault, as shown in Fig. 18.9.8.

The CT should have a knee point voltage equal to twice the relay setting voltage, as shown in Fig. 18.9.8. I_M is the magnetizing current taken by the CT at the setting voltage, N is the CT turns ratio, I_m is the current taken by the voltage-limiting device, Metrosil resistance (nonlinear resistance) at voltage V_s, I_{SR} is the current taken by the fault setting resistor at the setting voltage, $n =$ number of CTs in the busbar protection scheme, and I_v is the current taken by the supervision relay at setting voltage.

I_{op} is the relay operating current and should be at least 30% of the minimum fault current to insure relay operation.

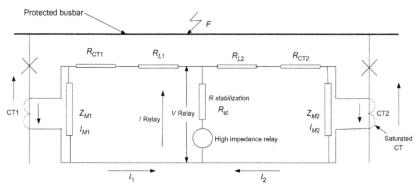

FIGURE 18.9.8 Internal fault in high-impedance differential protection relay.

18.9.3.3.2.1 Through-Fault Stability Limit

The relay has a stability limit, which is the maximum through-fault current under which the relay will remain stable.

18.9.3.3.3 Fault Setting Resistor

This fault setting resistor is used to increase the effective primary fault current by creating a shunt resistance across the relay circuit, as shown in Fig. 18.9.9.

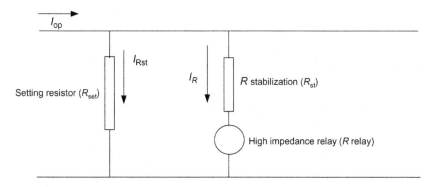

Where $I_{op} = I_R + I_{Rst}$

FIGURE 18.9.9 Fault setting resistor (R_{st}).

18.9.3.3.4 Check Zone Feature

As backup protection to confirm the tripping decision, a duplicate of the primary protection using a second set of CT on all circuits other than the bus section and coupler units is provided. The check system is arranged in a similar manner to the primary protection but forms one zone only, covering the whole of the busbars and does not discriminate between faults in busbar sections.

18.9.3.3.5   Nonlinear Resistance (Metrosils Resistance)

During internal faults of busbars, the high-impedance relay circuit represents a high burden to CTs, which leads to high voltage on the relay circuit which can damage the relay, so we use a nonlinear resistance called Metrosil resistance connected in parallel with the relay circuit as shown in Fig. 18.9.10, to limit this voltage.

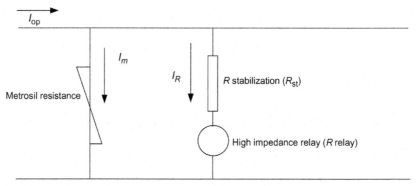

FIGURE 18.9.10   Metrosil (nonlinear resistor) of voltage relay in high-impedance differential protection scheme.

18.9.3.3.6   Busbar Protection Supervision Relay

When a CT is open-circuited the resultant unbalanced current in the busbar scheme will flow through the parallel combination of relay, metrosil, fault setting resistor, and CT magnetizing impedance—this causes the busbar protection to operate for load or through-fault conditions depending on the effective primary setting. We use a sensitive voltage operated relay to operate when there is unbalanced current equal to 10% of the least loaded feeder connected to the busbars, as shown in Fig. 18.9.11.

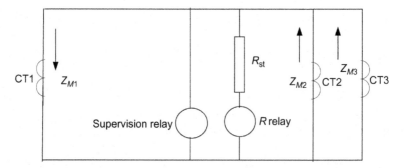

FIGURE 18.9.11   Supervision relay in high-impedance differential relay busbar protection scheme.

When there is an open circuit in CT1 then the current I_1 of CT1 will pass through the magnetizing impedance Z_{M2} and Z_{M3} and the relay total resistance (R relay + R_{st})

V supervision = I_1 ((R relay + R_{st}) // Z_{M2} // Z_{M3})

The action of this supervision relay after a time delay of about 3 seconds is to give an alarm that the busbar protection is faulty and to short circuit the buswires to prevent damage to the relay and stabilizing resistors due to the open circuit condition on one of the busbar protection CTs.

18.9.4   TYPICAL BUSBAR PROTECTION OF HIGH-IMPEDANCE TYPE

Refer to the AC and DC circuits of the two section single busbar scheme with one relay per zone circuits in Figs. 18.9.12 and 18.9.13.

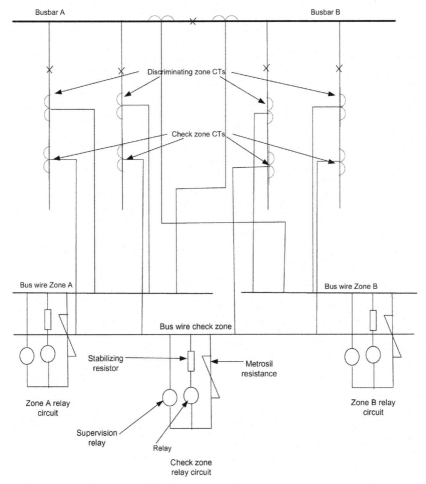

FIGURE 18.9.12 Simple example for a single line AC circuit of two section single busbar scheme with one relay per zone circuit.

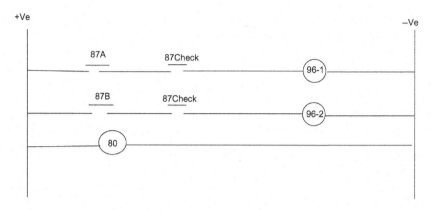

Where:

87A High impedance differential relay trip contact for Zone A

87A High impedance differential relay trip contact for Zone B

87 Check high impedance differential relay trip contact

80 DC voltage supervision relay

96 Trip relay

FIGURE 18.9.13 Simple example for DC circuit of two section single busbar scheme with one relay per zone circuit.

As shown in Fig. 18.9.12, the primary discriminating zone can discriminate the fault in each section of busbars A or B, but the check zone is protecting the whole system A and B as one connected busbar.

Subchapter 18.10

Unit Protections

18.10.1 INTRODUCTION

In this type of protection the protected zone is defined between two current transformers, for example, line differential protection and transformer differential protection.

This protection depends on a comparison between the current input to the protected zone and the current output from it. Refer Figs. 18.10.1 and 18.10.2 for the balanced current differential protection. A resistance is added in series with the relay circuit to balance the unbalanced currents; this resistor is called the stabilizing resistance.

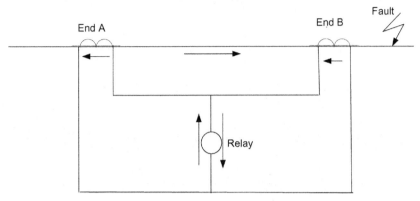

FIGURE 18.10.1   Balanced circulating current system, external fault—relay should not operate.

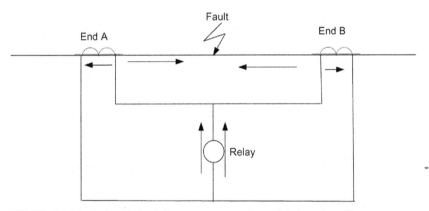

FIGURE 18.10.2   Balanced circulating current system, internal fault—relay will operate.

There is another differential relay scheme as balanced circulating current scheme is known as voltage balance differential relay. Refer Figs. 18.10.3 and 18.10.4.

The voltage balance differential scheme is used in feeder protection, but current balance differential scheme is used for generator, transformer, and switchgear main protection.

The principal of bias is applied to circulating current scheme to insure the stability of the differential relay under external (through) fault condition, then the relay will be driven to operate by the operating quantity and breaking by the bias quantity.

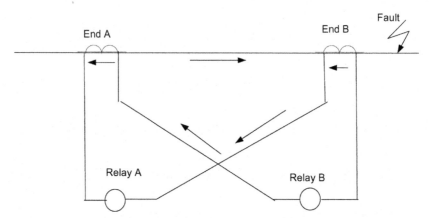

FIGURE 18.10.3 Balanced voltage system—external fault—relay should not operate.

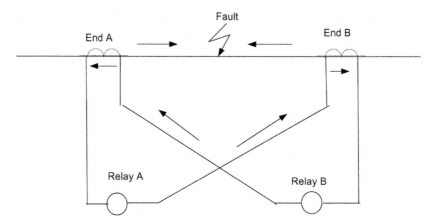

FIGURE 18.10.4 Balanced voltage system, internal fault—relay will operate.

18.10.2 APPLICATION OF DIFFERENTIAL PROTECTION

1. Generator differential protection;
2. Transformer differential protection;
3. Switchgear-busbar differential protection;
4. Feeder pilot-wire protection.

18.10.3 DIFFERENTIAL PROTECTION OF HIGH-VOLTAGE TRANSMISSION LINES (PILOT-WIRE PROTECTION)

In this type of protection that use the principal of balanced circulating current scheme to compare the currents in the two ends of the high-voltage line as shown in Fig. 18.10.5.

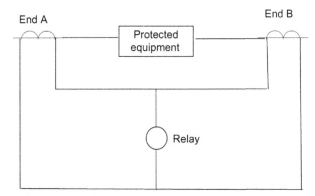

FIGURE 18.10.5 Merz-price differential protection.

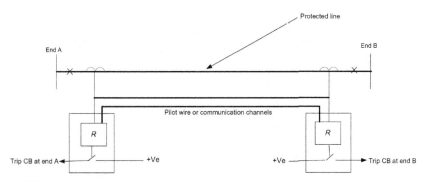

FIGURE 18.10.6 Practical pilot wire differential protection.

Refer to the unit pilot-wire differential protection in Fig. 18.10.6.

There are different methods to transfer the current signals over the line as follows:

Pilot wires;
Power line carrier (PLC);
Radio link;
Optical fiber; and
Pilot wires.

This system can be used for short lines less than 10 km.

18.10.4 POWER LINE CARRIER

This system is used to transfer the tripping and protection acceleration signal to the remote end of the line along with the voice signal. Refer Fig. 18.10.7.

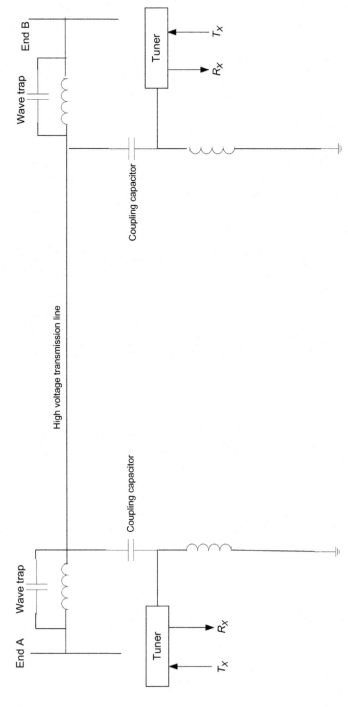

FIGURE 18.10.7 Communication through a high-frequency signal, interposed at the high-voltage line (power line carrier).

Radio links;
Few used in the past;
Optical fibers.

The most recent system is to transfer the current signals and all tripping and protection accelerating signals along of fiber optic between the two ends of the line. Refer Fig. 18.10.8.

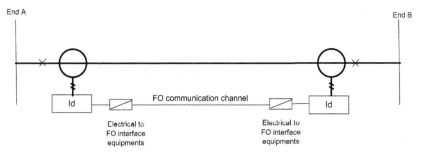

FIGURE 18.10.8 Line differential protection with a fiber-optic communication channel.

In recent relays there are two bias constant zones in its operating characteristic, the first one for low fault currents that need less bias and the second one for high fault currents that need more bias as shown in Fig. 18.10.9.

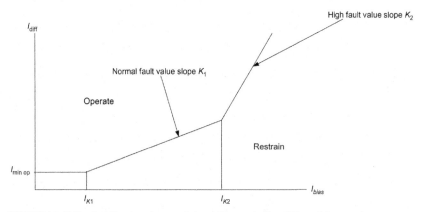

FIGURE 18.10.9 Stabilization characteristic of fiber-optic line differential protection.

Subchapter 18.11

Backup Protections

18.11.1 INTRODUCTION

The basic design of protective relays of a power system is designed as follows.

All elements of the power system are protected by at least one sensitive relay called a primary relay, which trips as quickly as possible, in the order of 100 ms when a fault occurs, as shown in Fig. 18.11.1.

These elements of the power system are protected by another set of relays, which are called backup protection, in the event of failure of the primary protection, as shown in Fig. 18.11.2.

Backup protection is divided into two types as follows:

1. Remote backup relays are located one bus away from the primary protection;
2. Local backup relays are at the same location as the primary protection.

18.11.2 REMOTE BACKUP PROTECTION

As shown in Fig. 18.11.2, the primary protection will trip breaker 1 and breaker 2 for a fault on line A instantaneously, but if the breaker 1 is not tripped the backup protection at breaker 3 will trip breaker 3, but with a time delay usually of 300 ms, the same applies if breaker 2 does not trip the backup protection at breaker 6, which will trip breaker 6 but with a time delay of 300 ms, this give a second protection to clear the fault with a time delay if the primary protection did not clear it instantaneously.

18.11.3 LOCAL BACKUP AND BREAKER FAILURE PROTECTION

In the case of relay failure, two sets of protection are installed at the local ends of the protected line. One is the primary and the second is the backup, alternatively these are called main 1 and main 2, in the case of one set failing to trip the breaker, the other relay trips. This normally utilizes two relays from different manufacturers, and if these need a communication channel they use separate communication channels for each relay.

However, in the case of a breaker failing to trip we use a BF scheme, as shown in Fig. 18.11.3.

When a fault occurs on the line controlled by breaker 1, the distance relay at the breaker 1 end sends a trip signal to breaker 1 at the same time it will send a start signal to the BF relay then The timer (T) will starts and the

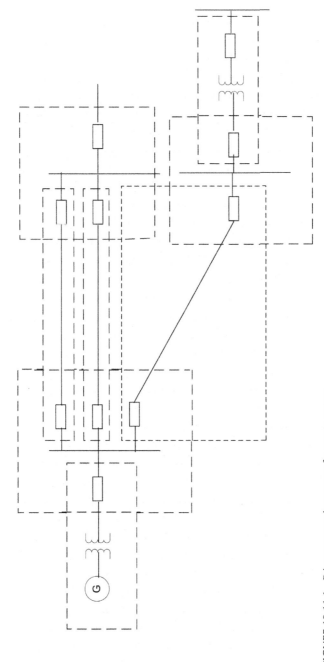

FIGURE 18.11.1 Primary protection zones of a power system.

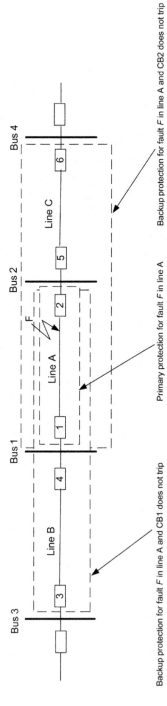

FIGURE 18.11.2 Remote backup protection zones of a power system.

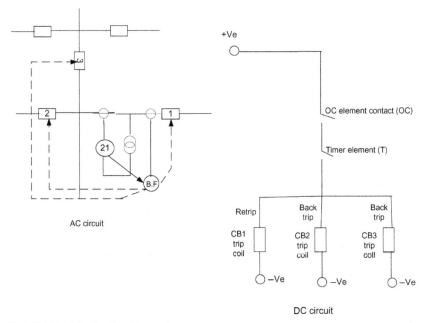

FIGURE 18.11.3   Breaker failure scheme.

current detector contact (OC) closes if the current is still passing, which means that when breaker 1 fails to trip after time T expires, the BF relay also sends a retrip to breaker 1 and a trip signal to trip breaker 2 and breaker 3 to stop the fault current flow and send a trip signal to the remote end of the line if this end feeds the fault.

BF relay time setting is set in the range of 110−180 ms, some utilities are at set 125 ms and others at 140 ms. This will be explained in the BF setting chapter of this book (Sub-chapter 19.7.3).

The total fault clearing time of BF relay is about 260−300 ms, as explained in the BF setting chapter of this book (Sub-chapter 19.7.3).

Subchapter 18.12

Transformer Protection

18.12.1   INTRODUCTION

Transformer faults can be classified as follows:

Winding and terminal faults;
Core faults;

Abnormal operating conditions, such as over voltages, over excitation, or overloading;
Uncleared external faults;
Transformer connections.

Group 1	0 degrees	Phase displacement	Y_{y0}, D_{d0}, Z_{d0}
Group 2	180 degrees	Phase displacement	Y_{d6}, D_{d6}, D_{z6}
Group 3	30 degrees	Lag phase displacement	Y_{d1}, D_{y1}, Y_{z1}
Group 4	30 degrees	Lead phase displacement	$Y_{d11}, D_{y11}, Y_{z11}$

Differential protection is not enough to protect the transformer against earth faults, especially in solid-earthed systems, and there is a need to have separate earth fault protection.

18.12.1.1   Transformer Inrush Current

When the transformer is first energized a transient current follows, which may reach instantaneous peaks of 8−30 times the transformer full load current.

Factors affecting the duration and magnitude of inrush current are as follows:

- Size of the transformer bank;
- Size of the power system;
- Resistance in the power system from the source to the transformer bank;
- Residual flux level;
- Type of iron used for the core and its saturation level.

There are three conditions which can produce a magnetizing inrush effect:

1. First energization;
2. Voltage recovery following external fault clearance;
3. Inrush current due to a parallel transformer being energized.

Inrush current can be detected by the second harmonic contents, which are high in inrush current and low in fault current so we can detect the second harmonic in inrush current and block the operation of differential protection at the starting of transformer energization.

Another method used to detect the inrush current is the zero detection in current waveform as the inrush current has a significant period in each cycle where the current is substantially zero, but the fault current passes through the zero point very quickly.

18.12.2   TRANSFORMER OVERLOAD PROTECTION

Excessive overloading will result in deterioration of insulation and after that lead to failure. Transformer normal operating temperature is 98°C and is equal to the transformer winding and cooling medium temperature of 78°C plus an ambient temperature of 20°C. This is the normal operating temperature, which does not affect the transformer service life. By using a thermal replica impedance relay incorporated with CT input the transformer, which has the same time constant as the transformer winding, can be protected against overloading condition.

18.12.3   TRANSFORMER OVERCURRENT PROTECTION

Applied to small transformers alone, and used as backup protection for large transformers, both instantaneous and time-delayed OC relays can be applied.

Inverse time OC relays are applied on the high voltage (HV) side of transformers and graded with those on the low-voltage (LV) side, which in turn must be graded with LV outgoing circuits.

High-set overcurrent relays can be used and set at 120%−130% of the through-fault level of the transformer.

Two stages of OC are used, the first stage trips the LVCB and the second stage trips the HVCB, as shown in Fig. 18.12.1.

An instantaneous relay (short circuit element) is used to provide a fast trip for the HV side.

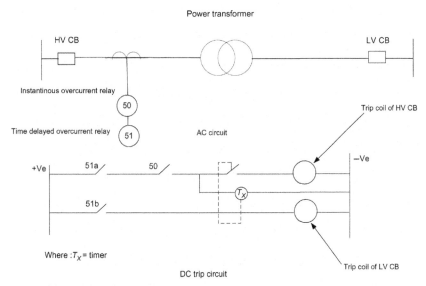

FIGURE 18.12.1   Overcurrent relays for transformers.

18.12.4 TRANSFORMER EARTH FAULT PROTECTION

Restricted earth fault protection is used to provide instantaneous earth fault protection to the transformer, utilizing high-impedance principal. As shown in Fig. 18.12.2 below, balanced earth fault for a delta (or unearthed star) winding can be provided by connecting three line CTs in parallel (residual connection). The relay will only operate for internal earth faults since the transformer itself cannot supply zero-sequence current to the system. The transformer must obviously be connected to an earth source, as shown in Fig. 18.12.2.

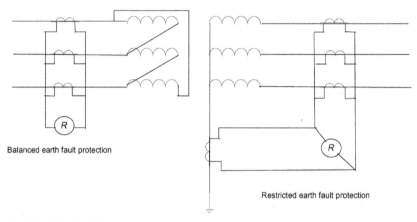

Balanced earth fault protection

Restricted earth fault protection

FIGURE 18.12.2 REF protection for transformers in star connection and balanced EF protection for transformers with delta connection.

On resistance-earthed power systems, unrestricted earth fault protection is referred to a standby earth fault protection of two stages of type inverse time relay, which match the thermal characteristic of the earthing resistor.

In practice, two EF relays with stage one set at 15% of full load current of transformer with a 5 seconds time delay and the send stage set to 20% of full load current of the transformer with an 8 seconds time delay, the first stage trips the LVCB and the send stage trips the HVCB.

Typical settings for REF protection are:

Solidly earthed systems	10%−60% of winding rated current
Resistance-earthed systems	10%−25% of minimum earth fault current for fault at transformer terminals

18.12.5 BUCHHOLZ PROTECTION

All types of fault within a transformer will produce heat, which will cause decomposition of the transformer oil. The resulting gases that are formed rise to the top of the tank and then to the conservator. A Buchholz relay connected between the tank and conservator collects the gas and gives an alarm when a certain volume of gas has been collected. A severe fault causes so much gas to be produced that pressure is built up in the tank and causes a surge of oil. The Buchholz relay will also detect these oil surges and under these conditions is arranged to trip the transformer CBs. The main advantage of the Buchholz relay is that it will detect incipient faults, which would not otherwise be detected by conventional protection arrangements. The relay is often the only way of detecting interturn faults, which cause a large current to flow in the shorted turns but due to the large ratio between the shorted turns and the rest of the winding, the change in terminal currents is very small.

18.12.6   DIFFERENTIAL PROTECTION

The function of differential protection is to provide faster and more discriminative phase fault protection than that obtainable from simple overcurrent relays. CTs on the HV side are balanced against CTs on the LV side. There are a number of different connections but there are some important points that are applicable to all schemes.

18.12.6.1   Transformer Connections

Consider a delta/star transformer. An external earth fault on the star side will result in zero-sequence current flowing in the line but due to the effect of the delta winding there will be no zero-sequence current in the line associated with the delta winding. In order to ensure stability of the protection this zero-sequence current must be eliminated from the secondary connections on the star side of the transformer, that is, the CTs on the star side of the transformer should be connected in delta. With the CTs on the delta side of the transformer connected in star, the 30 phase shift across the transformer is also catered for. Since the majority of faults are caused by flashovers at the transformer bushings, it is advantageous to locate the CTs in the adjacent switchgear.

There are a general rules for CT connections:

1. CT connections opposite to main power transformer connections, that is, star CTs on delta side and delta CTs on star side;
2. If similar primary terminals, that is, P1 or P2, are toward the transformer, then delta and star connection for the CTs should be the same as for the transformer (or 180 degrees opposite);
3. It is usual to assume that if current flows from R1 toward R2 then the secondary current will flow from a2 towards a1 (refer to Fig. 18.12.3).

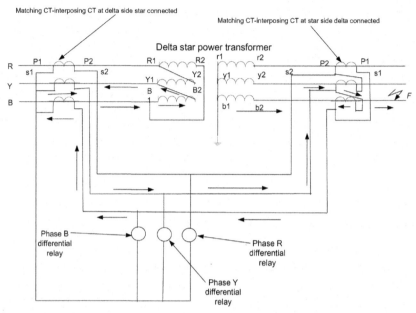

FIGURE 18.12.3 Delta star power transformer differential protection connections with earth fault on the low-voltage side on phase B.

18.12.6.2 Tap Changers

Any differential scheme can only be balanced at one point of the tap changer, and it is usual to choose CT ratios that match at the midpoint of the tap range. Note that this might not necessarily be the normal rated voltage. For example, if the tapping range is +10%, −20%, then the CT ratio should be based on a current corresponding to the −5% tap. The theoretical maximum out of balance in the differential circuit is then +15%.

18.12.6.3 Auto-Transformer Differential Protection

Auto-transformers can be adequately protected by high-impedance differential relays. The setting voltage is calculated assuming the worst through-fault conditions and therefore bias is not necessary. CTs are required on the HV, LV, and neutral connections and all must be the same ratio. The CT ratio should be based on the LV full load current for the nearest standard ratio. This form of protection will not detect interturn faults (refer to Fig. 18.12.4).

The corner of an unloaded delta tertiary winding can be earthed on the main transformer earth, thereby including the delta winding within the zone of protection. However, the percentage of the delta winding that is protected

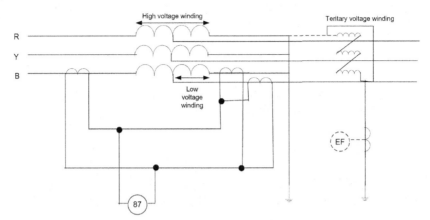

FIGURE 18.12.4   Auto-transformer differential protection.

will be minimal due to the high CT ratios that the protection requires. If more sensitive earth fault protection is required for the delta winding, a separate earth should be used, together with a suitable CT and relay. The protection is also stable under over fluxing conditions, and magnetizing inrush. If CTs are not available on the neutral, a biased differential relay can be used in the conventional way. The line CTs should be connected in delta (as for the star/star transformer). The biased differential relay will give a measure of interturn protection and will also detect delta tertiary faults.

18.12.6.4   Earthing Transformer Protection

If the earthing transformer is inside the zone of differential protection then no separate protection is required. Overcurrent protection can be provided by three overcurrent units connected to delta-connected main CTs, as shown in Fig. 18.12.5.

The protection will not operate on zero-sequence currents due to system earth faults because of the CT delta connection.

18.12.7   ZERO-SEQUENCE CIRCUITS OF TRANSFORMERS

18.12.7.1   Two-Winding Transformers

Consider, for example, a two-winding transformers delta/star connection, as shown in Fig. 18.12.6.

The zero-sequence currents will be there in the star connection side of the transformer but will not be there in line with the connection of the delta side (primary side) of the transformer.

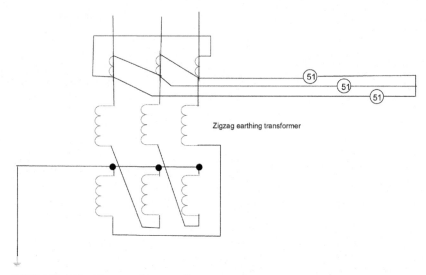

FIGURE 18.12.5 Zigzag earthing transformer protection.

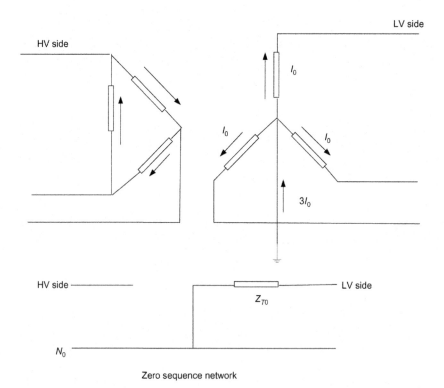

Zero sequence network

FIGURE 18.12.6 Equivalent zero-sequence network of a delta/star transformer.

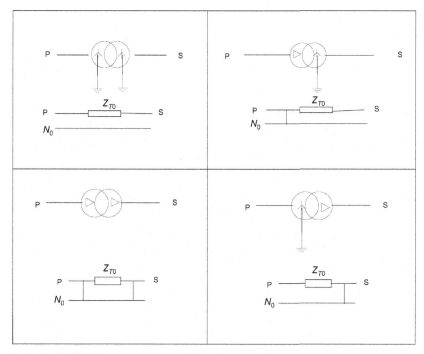

FIGURE 18.12.7   Zero-sequence network for different transformer connections.

Refer to Fig. 18.12.7 for the equivalent zero-sequence network of different transformer connections.

18.12.7.2   Three-Winding Transformers

Refer to Fig. 18.12.8 for the equivalent zero-sequence network of a star/star/delta transformer.

18.12.7.3   Zigzag Earthing Transformers

Refer to Fig. 18.12.9 for the zero-sequence network of a zigzag earthing transformer.

18.12.8   SYSTEM EARTHING

There are different methods of earthing the power system as follows:

1. Solid earthing:
    Direct connection for the star-connected system has the advantage of low overvoltages but the disadvantage is a high fault current used in

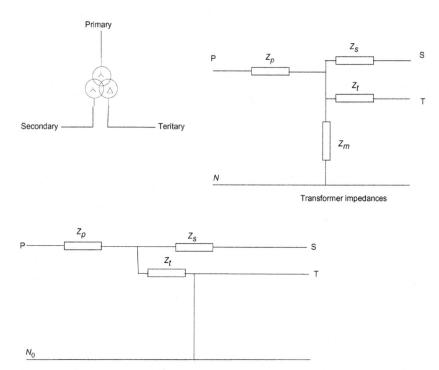

Transformer impedances

Zero sequence network of three-winding transformer star/star/delta connection

FIGURE 18.12.8   Zero-sequence network of a three-winding transformer.

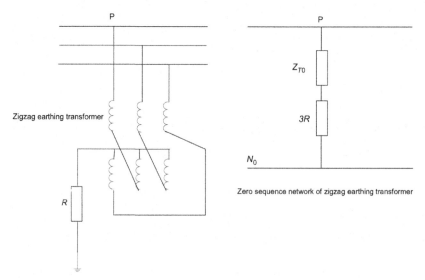

Zero sequence network of zigzag earthing transformer

FIGURE 18.12.9   Zero-sequence network of zigzag earthing transformer.

low-voltage systems and up to 660 V systems, as well as in high-voltage systems of more than 33 kV.

2. Resistance earthing:

   Connection to earth through a resistance for the star-connected system has the advantage of low transient overvoltages and low fault current used in generators and in systems of 660−33 kV when the fault current does not exceed the load current.

3. Reactance earthing:

   Connection to earth through a reactance for the star-connected system has the advantage of low fault current but a disadvantage of high transient overvoltages used for 660−33 kV systems when the single phase to earth fault current is less than the three-phase fault current.

4. Petersen coil earthing:

   Connection to earth through an adjustable reactance to compensate the capacitive earth current in the system for the star-connected system is good for transient faults.

5. Unearthed systems:

   These have the disadvantage of high transient overvoltages and the fault current will be the capacitive current.

18.12.9 COMPARISON OF HIGH-IMPEDANCE AND LOW-IMPEDANCE DIFFERENTIAL PROTECTION

High impedance has greater stability during CT saturation of one of the CTs for through faults than low impedance, but needs an identical CT of Class X and cannot share the same core with other protective relays. Mainly used in transformer (resistance earthed) REF protection schemes, they are low-cost, simple design, easy testing, and need accurate CT and wiring data to set the voltage setting. However, low impedance has low stability for through faults. It is mainly used in transformer protection but has the advantage of the ability to share the same core of the CT with other protective relays.

18.12.10 EXAMPLES OF TRANSFORMER DIFFERENTIAL PROTECTION CONNECTIONS

A D_{y11} transformer is illustrated in Fig. 18.12.10.

An auto-transformer Y_{y0} is illustrated in Fig. 18.12.11.

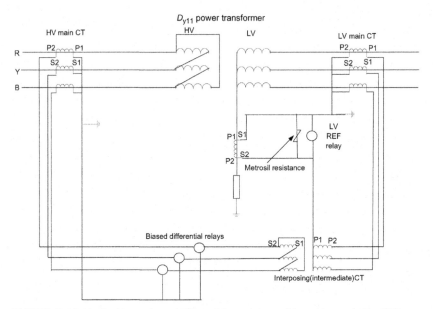

FIGURE 18.12.10 Dy11 transformer differential protection and low-voltage winding REF protection connections.

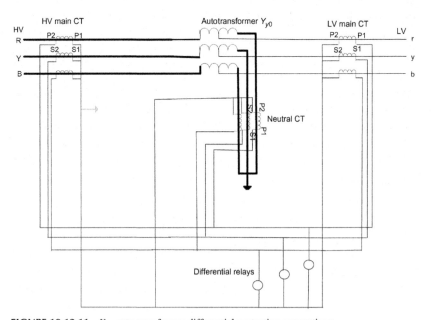

FIGURE 18.12.11 Y_{y0} auto-transformer differential protection connections.

Chapter 19

Protection Relays Settings

INTRODUCTION

This chapter introduces relay settings for different types of relays used in power system, with real worked examples.

Subchapter 19.1

Overcurrent Protection Settings

19.1.1 INTRODUCTION

Overcurrent protection is used in extra-high voltages and high-voltages networks as the main protection, and in medium networks it is also used as backup protection for power transformers.

Time grading for overcurrent relays is used in the radial system.

The minimum possible time delay is selected for the relay which protects the line farthest from the source. The grading time between the successive relays will be in the order of 0.4–0.5 seconds, and with modern relays it can be of the order of 0.35 seconds.

Current settings for an overcurrent relay should consider the following factors:

1. Maximum possible loading;
2. Transient currents due to switching of transformers, inrush current, heavy motor restarting currents after clearing external faults;
3. Drop off/pickup ratio, which is called the reset ratio.

$$I_{set} = (K \times R)/F \times I_{L\,Max}$$

where:

$K =$ safety factor $= 1.2$ $R =$ restarting factor, which depend on the load profile
Set to 2 for industrial loads

Set to 1.5 for residential loads;

F = reset ratio = 0.85 for electromechanical relays and set to 0.9–0.95 for static and digital relays;

$I_{L\ Max}$ = maximum expected load;

$I_{set} = 2$–$2.5\ I_{L\ Max}$.

The performance of overcurrent relays is checked by calculating the sensitivity factor, S, which is the ratio of the minimum short circuit current following at the relay in the end protected line in a radial system to the relay setting as follows:

$$S = I_{sc}\ \text{minimum}/I_{set}\ \text{should be} > 1.5$$

19.1.2 TRANSMISSION LINE OVERCURRENT PROTECTION

Fig. 19.1.1 illustrates the overcurrent protection of a high-voltage transmission line

The circuit data include:

Overcurrent relay type MCGG22.

Application:

Backup protection 500 kV, fault level at end B = 12,000 MVA, current setting of relay at A = 800 A.

$I_s = 800$ A   current transformer (CT) ratio $= \frac{1600}{1}$

$$I_s = 800\frac{1}{1600} = 0.5\text{A}$$

Plug setting multiplier (PSM)

$$\text{PSM} = \frac{I_f}{I_s} \rightarrow I_F = \frac{\text{MVA}_F}{\sqrt{3} \times V}$$

$$I_F = \frac{12,000}{\sqrt{3} \times 500} = 13.856\ \text{KA}$$

FIGURE 19.1.1 Overcurrent protection of a high-voltage transmission line.

$I_F = 13,856$ A
$I_S = 800$ A

$$\text{PSM} = \frac{13,856}{800} = 17.32$$

Using a standard inverse curve of the relay then we have the relay operating time T_R:

$$T_R = \frac{0.14}{\text{PSM}^{0.02-1}} = \frac{0.14}{17.32^{0.02-1}}$$

$T_R = 2.385$
Assuming that the relay should operates on 1 second then we have:
t.operation $= \text{TM}.T_R$
Where TM = time multiplier

$$\text{TM} = \frac{1}{2.385} = 0.419$$

Relay (A) setting:
$I_{\text{set}} = 800$ A (Primary)
$I_{\text{set secondary}} = 0.5$ A
$\text{PSM} = 17$
$\text{TM} = 0.4$
Operating time for a fault at end B will be:

$$t.\text{operation} = 0.4 \times \frac{0.14}{17^{0.02-1}}$$

t.operation $= 0.96$ seconds

19.1.3 TRANSFORMER OVERCURRENT PROTECTION

Transformer overcurrent protection is illustrated in Fig. 19.1.2.
Transformer data:
250 MVA   220/66 kV $Z_{\text{Tr}}\% = 10\%$ on MVAb = 100
Relay R_2 setting on low voltage (LV) feeder F_1:

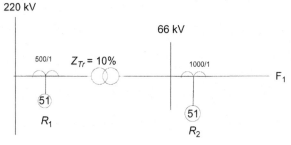

FIGURE 19.1.2 Transformer overcurrent protection.

$$MVA_F = \frac{MVA_b \times 100}{Z\%} = \frac{100 \times 100}{10} = 1000MVA$$

$$\text{If of HV side of Tr} = \frac{MVA_F}{\sqrt{3} \times 220} = \frac{1000}{\sqrt{3} \times 220} = 2624A$$

$$\text{IF of LV side of Tr} = \frac{1000}{\sqrt{3} \times 66} = 8747A$$

$I_{set} = 600$ A, CT $= 500/1$
TM $= 0.4$
PSM $= \frac{8747}{600} = 14.58$

$$\text{Relay operating time} = T_R = \frac{0.14}{14.58^{0.02-1}}$$

$T_R = 2.54$
For a fault of 8747 A the operating time will be:
$t.op = T_R.TM = 2.54 \times 0.4 = 1.016$ second
With a grading time step $= 0.4$ second.
Then the operating time of R_1 should be more than $= 1.016 + 0.4 = 1.416$ second.
Relay R_1 setting on the HV side of Tr
I_F on HV side $= 2624$ A
We can set the relay as follows: $I_{set} = 1.45 \times I_{Tr}$ HV side

$$I_{Tr}\text{HV side} = \frac{MVA}{\sqrt{3} \times VHV} = \frac{250 \times 10^3}{\sqrt{3} \times 220} = 656A$$

$I_{set} = 1.45 \times 656 = 951.2$ A

$$PSM = \frac{I_{Fault}}{I_{set}} = \frac{2624}{951.2} = 2.76$$

Set top $= 1.5$ second, as seen above it should be more than 1.416 second.
Top $= TM.T_R$

$$T_R = \frac{0.14}{2.76^{0.02-1}} \text{ SI} - \text{Curve}$$

TR $= 6.825$

$$TM = \frac{top}{T_R} = \frac{1.5}{6.825} = 0.219$$

R_2 Setting
$I_{set} = 600$ A   PSM $= 14.58$ TM $= 0.4$ CT $= 500/1$
R_1 Setting
$I_{set} = 951.2$ A   PSM $= 2.76$ TM $= 0.219$ CT $= 1000/1$

Subchapter 19.2

Feeder Backup Overcurrent Protections

19.2.1 INTRODUCTION

Overcurrent protection here is used as backup protection, with a considerable time delay, which means it works as a second line of defense to protect the line if the main or primary protection fails to isolate the line for the fault instantaneously.

19.2.2 WORKED EXAMPLE

Fig. 19.2.1 gives an example of line backup overcurrent protection.
Line AB 500 kV maximum load current $= I_{L\ Max} = 1200$ A
For OC relay at A (R):
$I_{set} = 1.2 \times I_{L\ Max}$
$= 1.2 \times \frac{1200}{1000} = 1.44$ A
PSM calculation:

$$\text{PSM} = \frac{I_F}{I_S} \text{ and } I_F = \frac{\text{MVA}_F}{\sqrt{3} \times V}$$

$$I_F = 80 \text{ KA}$$
$$I_S = 1200 \text{ A}$$
$$\text{PSM} = \frac{80,000}{1000} = 80$$

Using a standard inverse curve of the OC relay we then have:

$$T_R = \frac{0.14}{\text{PSM}^{0.02-1}} = \frac{0.14}{80^{0.02-1}} = 1.528$$

where T_R = relay operation time assuming that the relay will operates on 1 second for this fault then $t.\text{operation} = \text{TM}.T_R$, where TM = time multiplier.

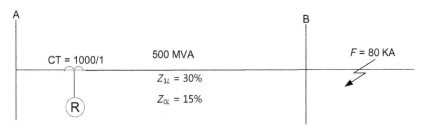

FIGURE 19.2.1   Line backup overcurrent protection.

$$\text{Then TM} = \frac{t.\text{operation}}{T_R} = \frac{1}{1.528} = 0.654$$

Relay setting:
$I_{set} = 1440$ A
$= 1.44$ A (secondary)
PSM $= 80$ TM $= 0.65$
Operation time at a fault after end B will be:

$$t = 0.65 \frac{0.14}{80^{0.02-1}} = 0.993s$$

For an earth fault relay at end A connected to the residual connection of CTs at end A:

$$\%Z_S = \frac{\text{MVA}_b \times 100}{\text{MVA}_F} = \frac{100 \times 100}{80 \times \sqrt{3} \times 500} = 0.144$$

For a single phase fault at end B we draw the sequence network diagram as shown in Fig. 19.2.2.
PSM calculation:
$I_F = 29.3$ KA $I_{set} = 800$ A

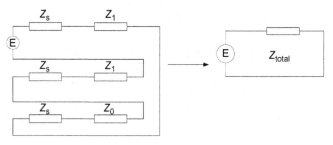

FIGURE 19.2.2   Sequence networks.

$$Z_t = 3Z_s + 2Z_1 + Z_0$$
$$= 3 \times 0.144 + 2 \times 0.3 + 0.15$$
$$= 0.432 + 0.6 + 0.15 = 1.182\%$$

$$I_0 = \frac{100.100}{1.182 \times \sqrt{3} \times 500} = 9.769 \text{ KA}$$

$$I_F = 3I_0 = 3 \times 9.769 = 29.3 \text{ KA}$$
$$I_{set} = 800 \text{ A}$$
$$I_{set} = \frac{800}{1000} = 0.8 \text{ A (secondary)}$$

$$\text{PSM} = \frac{I_F}{I_S} = \frac{29.3 \times 10^3}{800} = 36.625$$

Assuming that the relay will operates on 1 second for a fault on end B:

$$\text{TM} = \frac{t.\text{operation}}{T_R}$$

$$T_R = \frac{0.14}{\text{PSM}^{0.02-1}} = \frac{0.14}{36^{0.02-1}} = 1.884$$

$$\text{TM} = \frac{1}{1.884} = 0.530$$

Relay setting:
$I_{\text{set}} = 800$ A
$= 0.8$ A secondary
PSM $= 36$ TM $= 0.53$
$t.\text{operation} - 0.53 \dfrac{0.14}{36^{0.02-1}}$
$= 0.998$ second

Subchapter 19.3

Feeder Unit Differential Protection

19.3.1 INTRODUCTION

This protection is used to protect overhead transmission lines and is commonly used as main 2 in utilities. The disadvantage of this protection is the blind zone, which is not protected by this protection—this means the zone between the circuit breaker (CB) and the current transformer in one end—as this zone is protected by backup protections (Fig. 19.3.1).

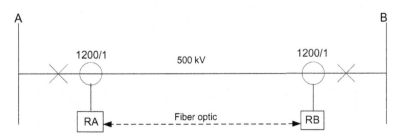

FIGURE 19.3.1 Line differential protection.

19.3.2 WORKED EXAMPLE

Worked example:

Line A−B 500 kV B = 1.877 CT = 1200/1 at both sides choose the relay characteristic shown in Fig. 19.3.2

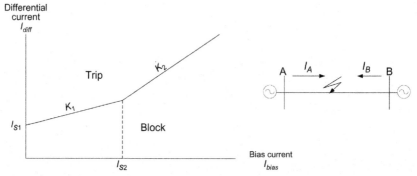

FIGURE 19.3.2 Line differential protection characteristic.

$$|I_{\text{diff}}| = |I_A + I_B|$$
$$|I_{\text{bias}}| = 0.5[|I_{A1} + I_{B1}|]$$
$$\text{Line charging current} = \frac{B \times 1000}{V \times \sqrt{3}}$$
$$I_{\text{ch}} = \frac{1.877 \times 1000}{500 \times \sqrt{3}}$$

$$= 2.167 \text{ A}$$

$$I_{\text{set min}} = 2.5 \times 2.167 = 5.4175 \text{ A}$$

$$I_{\text{set}} = 0.2 \, I_N$$

$$I_{\text{set}} = 0.2 \times 1200 = 240 \text{ A}$$

$$I_{\text{set}} > 2.5 \, I_{\text{ch}}$$
$$I_{S1} = 240 \text{ A primary}$$

$$I_{S1} = \frac{240}{1200} = 0.2 \text{ A}$$

$$I_{S2} = 2I_N$$
$$= 2 \times 1200 = 2400 \text{ A}$$

$$I_{S2} = 2400 \text{ A primary}$$
$$I_{S2} = \frac{2400}{1200} = 2 \text{ A secondary}$$

$$K_1 - \text{slop1} = 30 \, \%$$
$$K_2 - \text{slop2} = 150 \, \%$$

Subchapter 19.4

Setting of Distance Protection

19.4.1 INTRODUCTION

Distance protection is widely used as the main protection—primary protection—in overhead transmission lines and has the advantage that it gives backup protection for adjacent lines. However, it has the disadvantage that it does not have accurate measurements for close-up faults near the remote end of the line due to the errors in CT and PT measurements and the transient nature of the short circuit current, also due to the tower arc fault resistance—what we call distance protection overreaching and underreaching—that it is not accurate in measurement in short transmission lines due to the effect of the system impedance ratio.

19.4.2 WORKED EXAMPLES

Worked example 1

Most distance protection is set to have three zones and one backup start zone. The setting of the first zone is as follows for a relay at end A:

$Z_1 = 0.85\ Z_L$, $t_1 = 0$ instantaneously

where Z_L = line impedance.

Setting of the second step zone 2 is as follows for a relay at end A (Fig. 19.4.1):

The Z_2 setting is the lower value of the following:

$$Z_2 = 0.85\ (Z_{L1} + 0.85\ Z_{L2})$$
$$Z_2 = 0.85\ (Z_{L1} + Z_T)$$

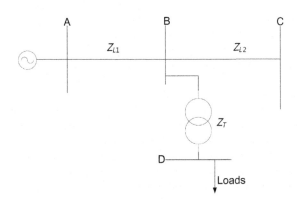

FIGURE 19.4.1 Distance protection application setting: Example 1.

Worked example 2

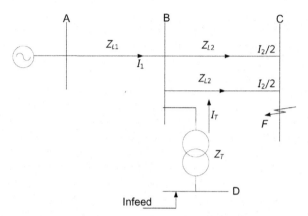

FIGURE 19.4.2 Distance protection application setting: Example 2.

For the following network configuration, the zone 2 setting will be as follows for a relay at end A:where $\frac{I_2}{I_1}$ is called the infeed factor "F"

$$Z_2 = 0.85 \left(Z_{L1} + \frac{I_2}{2I_1} \times 0.85 Z_{L2} \right)$$

$$Z_2 = 0.85 \left(Z_{L1} + \frac{I_T}{I_1} Z_T \right)$$

In general:

$$\frac{Z_2}{Z_{L1}} \geq 1.2 \rightarrow Z_2 = 120\% \ Z_{L1}$$

t_2 is set to be 0.5 second.

The setting of zone 3 in general is:

$$\frac{Z_3}{Z_{L1}} \geq 1.5 \ Z_3 = 150\% \ Z_{L1}$$

t_3 is set to be 0.9 second.

Final (Backup Step)

The current starter is used for simple networks where the minimum short circuit current is higher than the maximum load currents.

An impedance starter is more commonly used as this start impedance stage should not work in the load area and also should not work for the transformer low-voltage faults.

The lower value of Z_{st} should be chosen.

1. $Z_{st} \leq 0.5\ Z_{L\ min}$; this means less than half of the minimum load impedance;

2. $Z_{st} \geq 1.25\ Z_3$.

$t_{st} = 1.3$ second.

The Compensation Factor (K_0)

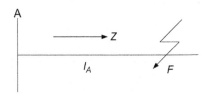

FIGURE 19.4.3   Distance protection compensation factor (K_0) for earth fault measurement.

The voltage measured by the relay at A is:

$$V_{RA} = Z_1 \left(I_A + 3I_0 \frac{Z_0 - Z_1}{3Z_1} \right)$$

$$I_R = I_A + K_0 3I_0$$

$$\text{Where } K_0 = \left(\frac{z_0 - z_1}{3z_1} \right)$$

The compensation factor is set in the relay by the values of Z_1 and Z_0 to have the correcting value of distance relay measure the line positive impedance, proportional to its length irrespective of the kind of fault (earth faults or phase faults).

Distance Protection Setting

$Z_{1F} = 80\%$ of the protected line;

$Z_{2F} = 100\%$ of the protected line + 50% of the shortest adjacent line;

$Z_{3F} = 100\%$ of the protected line + 125% of the longest adjacent line;

$Z_{3R} = 25\%$ of the protected line;

$F =$ forward direction, $R =$ reverse direction.

Worked example 3

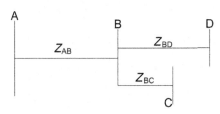

FIGURE 19.4.4   Distance protection application setting: Example 3.

Data:

AB = protected line length = 60 km 220 kV CT = 800/1 $PT = \frac{220,000}{100}$

$$Z_{1P} = 0.0412 + J\ 0.302\ \Omega\ /\ km \times 60$$

$$Z_{10} = 0.25 + J\ 0.95\ \Omega\ /\ km \times 60$$

BC = shortest line adjacent to the protected line AB length = 50.5 km

$$Z_{2P} = 0.0412 + J\ 0.302\ \Omega\ /\ km \times 50.5$$

BD = longest line adjacent to the protected line AB length = 68 km

$$Z_{3P} = 0.0412 + J\ 0.302\ \Omega\ /\ km \times 68$$

CT ratio = 800/1 = 800
Rated voltage = 220 kV
Rated current = 800 A
$PT = \frac{220,000}{100}$
Ratio = 2200

$$
\begin{aligned}
Z_p \text{ line AB} \quad &= 60(0.0412 + J\ 0.302) \\
&= 2.472 + J\ 18.12 \\
&= 18.287 \angle 82.231°\ \Omega
\end{aligned}
$$

$$
\begin{aligned}
Z_P \text{ line BC} \quad &= (0.0412 + J\ 0.302)50.5 \\
&= 2.0806 + J\ 15.251 \\
&= 15.356 \angle 82.231°
\end{aligned}
$$

$$
\begin{aligned}
Z_P \text{ line BD} \quad &= (0.0412 + J\ 0.302)68 \\
&= 2.8016 + J\ 20.536 \\
&= 20.726 \angle 82.231°
\end{aligned}
$$

$$
\begin{aligned}
Z_{1\ Reach} \quad &= 0.8\ Z_P \text{ line AB} \\
&= 0.8 \times 18.287 \angle 82.231° \\
&= 14.629 \angle 82.231°\ \Omega \\
&= 14.629\ \Omega
\end{aligned}
$$

$$
\begin{aligned}
Z_{2\ Reach} \quad &= Z_P \text{ line AB} + 0.5\ Z_P \text{ line BC} \\
&= (18.287 + 0.5 \times 15.356) \angle 82.231° \\
&= 25.965 \angle 82.231°\Omega \\
&= 25.965\ \Omega
\end{aligned}
$$

$$
\begin{aligned}
Z_{3\ Reach} \quad &Z_P \text{ line AB} + 1.25\ Z_P \text{ line BD} \\
&= (18.287 + 1.25 \times 20.726) \angle 82.231° \\
&= 44.1945 \angle 82.231° \\
&= 44.1942\ \Omega
\end{aligned}
$$

$$Z_3 \text{ reverse} = 0.25 \times Z_P \text{ line AB}$$
$$= 0.25 \times 18.287 \angle 82.231°$$
$$= 4.571 \angle 82.231° \ \Omega$$
$$= 4.571 \ \Omega$$

The earth fault residual compensation factor is calculated as:

$$K_0 = \frac{Z_{10} - Z_{1P}}{3Z_{1P}}$$

$$Z_0 \text{ line AB} = (0.25 + J\,0.95) \times 60$$
$$Z_0 \text{ line AB} = 15 + 57J = 58.94 \angle 75.25\,1°$$

$$K_0 = \frac{58.94 \angle 75.25 - 18.287 \angle 82.231}{3 \times 18.287 \angle 82.231}$$

$$= \frac{15 + J56.99 - 2.472 - J18.12}{54.861 \angle 82.231}$$

$$= \frac{12.528 + J38.87}{54.861 \angle 82.231}$$

$$= \frac{40.839 \angle 72.135°}{54.861 \angle 82.231}$$

$$K_0 = 0.7444 \angle -10.096$$
$$K_0 = 0.744$$

The setting secondary values are:

$$Z_1 \text{ Reach} = 14.629 \times \frac{\text{CTratio}}{\text{PTratio}}$$

$$= 14.629 \times \frac{800}{2200}$$

$$= 14.629 \times 0.363$$

$$= 5.319\Omega, \ t_1 \text{instantaneous}$$

$$Z_2 \text{ Reach} = 25.965 \times 0.363$$
$$= 9.425 \ \Omega \, , \ t_2 = 500 \text{ ms}$$

$$Z_3 \text{ Reach} = 44.1945 \times 0.363$$
$$= 16.042 \ \Omega \, , \ t_3 = 900 \text{ ms}$$

$$Z_3 \text{ Reverse} = 4.571 \times 0.363$$
$$= 1.659 \ \Omega$$

Power Swing Blocking Settings

$$Z_{PSB} = 1.3\, Z_{3\,F}$$
$$= 1.3 \times 16.042$$
$$= 20.854\ \Omega\ \text{(secondary value)}$$
$$Z_{PSB} = 20.854\ \Omega$$

where Z_{PSB} = power swing blocking impedance setting, Z_3 = relay third zone. Load encroachment into $Z_{3\ reach}$

$$Z_{Load\ min} = \frac{0.8\ \text{rated voltage}}{1.2\ \text{rated current}\ \sqrt{3}}$$
$$= \frac{0.8 \times 220000}{1.2 \times 800 \times \sqrt{3}}$$
$$= 105.85\ \Omega$$

$$Z_{Load\ min\ secondary} = 105.85 \times 0.363 = 38.422\ \Omega$$
$$Z_{3\ Reach} < 0.9\, Z_{Load\ min}$$

$16.041 < 0.9 \times 38.422$

$16.041 < 34.58$; the above criteria are correct.

Subchapter 19.5

Differential Protection Setting

19.5.1 INTRODUCTION

Transformer protection can be high-impedance differential protection or low-impedance differential protection. Both have setting calculations as shown in the following examples.

19.5.2 LOW-IMPEDANCE DIFFERENTIAL PROTECTION

19.5.2.1 Worked Example 1

(Fig. 19.5.1)

Tr 66/11 kV 25MVA Dy1

Tap changer = ± 10% of 11 kV side nominal voltage.

Calculation:

HV side:

$$I_1 = \frac{25 \times 10^3}{\sqrt{3} \times 66} = 219\ A$$

Choose CT ratio = 250/1

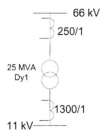

66 kV

250/1

25 MVA
Dy1

1300/1

11 kV

FIGURE 19.5.1   Transformer low-impedance differential protection.

$$i_1 = 219 \; \frac{1}{250} = 0.876 \text{ A}$$

where I_1 = primary current at the HV side of the transformer; i_1 = secondary current at the HV side of the transformer.

HV correction factor = $\frac{1}{i_1} = \frac{1}{0.876} = 1.145$

LV side:

$$I_1 = \frac{25 \times 10^3}{\sqrt{3} \times 11} = 1312 \text{ A}$$

Choose CT ratio = 1300/1

$$i_2 = 1312 \; \frac{1}{1300} = 1.009 \text{A}$$

where I_2 = primary current at the LV side of the power transformer; i_2 = secondary current at the LV side of the power transformer.

$$\text{LV correction factor} = \frac{1}{1.009} = 0.991$$

For phase correction and zero-sequence current filtering, we use interposing CT as follows:

HV side we use Yy_0 transformer
LV side we use Yd_{11} transformer

as shown in Fig. 19.5.2.

Now the current input to the relay is 1 A $\angle 0$ degree after the interposing CT ratio and phase correction at the HV side and the LV side of the power transformer.

Minimum Bias Current

The maximum voltage is calculated at the low-voltage side.

$$V_{\max} = 11 \times 1.1 = 12.1 \text{ kV}$$

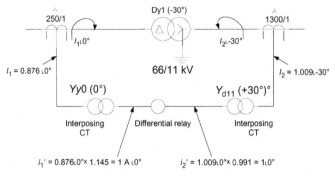

FIGURE 19.5.2 Interposing transformer in transformer differential protection.

$$I_2 = \frac{25 \times 10^3}{\sqrt{3} \times 12.1} = 1193 \text{ A}$$

$$I_R = \text{relay current} = 0.95 \times \frac{1193}{1300} = 0.871 \text{ A}$$

$$\text{Min Bias} = 2\left(I_{\text{Relay set Normal}} - I_{\text{Relay at max voltage}}\right)$$
$$= 2\left(1 - 0.871\right) = 0.256$$

The minimum voltage is calculated at the low-voltage side.

$$V_{\text{min}} = 11 \times 0.9 = 9.9 \text{ kV}$$

$$I_2 = \frac{25 \times 10^3}{\sqrt{3} \times 9.9} = 1458 \text{ A}$$

$$I_R = \text{relay current} = 0.95 \times \frac{1458}{1300} = 1.065 \text{ A}$$

$$\text{Min Bias} = 2\left(I_{\text{Relay at min voltage}} - I_{\text{Relay set at nominal voltage}}\right)$$
$$= 2\left(1.065 - 1\right) = 0.13$$

Set the minimum bias at 26%.

Minimum bias is the highest calculated value of bias at the maximum voltage and the minimum voltage at the low-voltage side of the power transformer.

Pickup or $I_{d\,\text{min}}$ set to 10% I_n

$$I_{d\,\text{min}} = 0.1 \times 1 = 0.1 \text{ A}$$

First slope bias = 26%
Second slope bias = 80% as shown in Fig. 19.5.3.

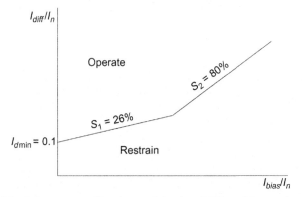

FIGURE 19.5.3 Relay–operate–bias characteristic of used differential protection.

19.5.2.2 Worked Example 2

A transformer with three-winding Y_{yd11} 500/220/11 kV
500 MVA/400 MVA/100 MVA
Tap changer $= \pm 10\%$ of 220 kV side and tap charger $= \pm 10\%$ of 11 kV side calculation.
HV side: 500 kV

$$I_1 = \frac{500 \times 10^3}{\sqrt{3} \times 500} = 577.35 \text{ A}$$

Choose CT $= 600/1$

$$i_1 = \frac{577.35}{600} = 0.962 \text{ A}$$

$$\text{HV correction factor} = \frac{1}{0.962} = 1.039$$

LV side: 220 kV side

$$I_2 = \frac{400 \times 10^3}{\sqrt{3} \times 220} = 1049.72 \text{ A}$$

Choose CT $= 1000/1$

$$i_2 = \frac{1049.72}{1000} = 1.0497 \text{ A}$$

$$\text{LV correction factor} = \frac{1}{1.0497} = 0.952$$

Tertiary side 11 kV side:

$$I_3 = \frac{100 \times 10^3}{\sqrt{3} \times 11} = 5248.63 \text{ A}$$

Choose CT = 5000/1

$$i_3 = \frac{5248.63}{5000} = 1.0497 \text{ A}$$

Tertiary side correction factor $= \frac{1}{1.0497} = 0.952$

For phase correction and zero $=$ sequence current filtering we use the following interposing "auxiliary" transformer for current as follows (refer to Fig. 19.5.4):

HV side: 500 kV Y_{d11} ICT;

LV side: 220 kV Y_{d11} ICT;

Tertiary side: 11 kV Y_{y0} ICT.

Refer to Fig. 19.5.5 for a detailed connection diagram of three-winding transformer-biased differential protection.

Minimum bias current (220 kV side)

Maximum voltage at the low-voltage side:

$$V_{max} = 220 \text{ kV} \times 1.1 = 242 \text{ kV}$$

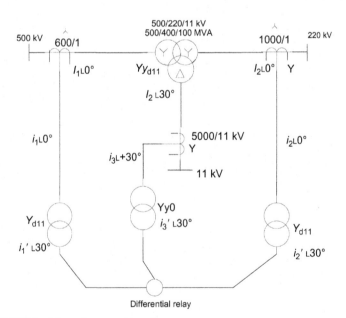

FIGURE 19.5.4 Interposing current transformers.

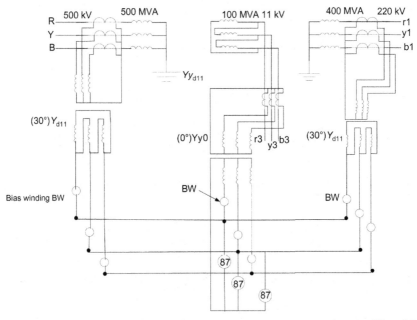

FIGURE 19.5.5 Detailed connection diagram of three-winding transformer differential protection.

$$I_2 = \frac{400 \times 10^3}{\sqrt{3} \times 242} = 954.3 \text{ A}$$

$$I_{\text{Relay}} = 0.95 \times \frac{954.3}{1000} = 0.906 \text{ A}$$

$$\text{Min Bias} = 2 \left(I_{R \text{ at nominal}} - I_{R \text{ at } V\text{max}} \right)$$
$$= 2 \left(1 - 0.906 \right) = 0.188$$

Minimum voltage at the low-voltage side:

$$V_{\text{min}} = 220 \text{ kV} \times 0.9 = 198 \text{ kV}$$

$$I_2 = \frac{400 \times 10^3}{\sqrt{3} \times 198} = 1166.4 \text{ A}$$

$$I_{R \text{ at min voltage}} = 0.95 \times \frac{1166.4}{1000} = 1.108 \text{ A}$$

$$\text{Min Bias} = 2 \left(I_{R \text{ at min voltage}} - I_{R \text{ at nominal voltage}} \right)$$
$$= 2 \left(1.108 - 1 \right) = 0.218$$

For the low-voltage side choose bias = 22%.

Minimum bias current for the 11 kV side with maximum voltage at the tertiary side is calculated as follows:

$$V_{max} = 11 \times 1.1 = 12.1 \text{ kV}$$

$$I_3 = \frac{100 \times 10^3}{\sqrt{3} \times 12.1} = 4771.5 \text{ A}$$

$$I_{R \text{ at } Vmax} = 0.95 \times \frac{4771.5}{5000} = 0.906$$

$$\text{Min Bias} \quad = 2 \left(I_{R \text{ at } Vnominal} - I_{R \text{ at } Vmax} \right)$$
$$= 2 \left(1 - 0.906 \right) = 0.188$$

Minimum voltage at the tertiary side (11 kV):

$$V_{min} = 11 \times 0.9 = 9.9 \text{ kV}$$

$$I_3 = \frac{100 \times 10^3}{\sqrt{3} \times 9.9} = 5832 \text{ A}$$

$$I_{R \text{ at } Vmin} = 0.95 \times \frac{5832}{5000} = 1.108 \text{ A}$$

$$\text{Min Bias} \quad = 2 \left(I_{R \text{ at } Vmin} - I_{R \text{ at } Vnominal} \right)$$
$$= 2 \left(1.108 - 1 \right) = 0.216$$

Choose bias = 22% for the 11 kV side.

We then choose a bias slop of 22% for the differential relay (as shown in Fig. 19.5.6):

$I_{\text{diff. min}} = 0.1 \text{ A}$

Note: second slope = 80%, to give more stability for the relay operation on external faults (through faults).

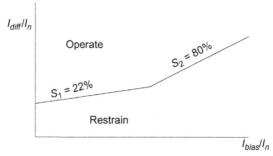

FIGURE 19.5.6 Relay−operate−bias characteristic of a three-winding transformer biased differential protection.

19.5.3 HIGH-IMPEDANCE DIFFERENTIAL PROTECTION SETTING

19.5.3.1 Worked Example 1 (Fig. 19.5.7)

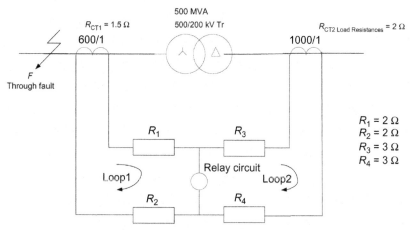

FIGURE 19.5.7 500/220 kV transformer Y_{d1} 500 MVA. Protected by high-impedance differential protection.

The data for this example include:

Lead resistance:

$R_1 = 2\ \Omega$ $R_2 = 2\ \Omega$ $R_3 = 3\ \Omega$ $R_4 = 3\ \Omega$

$R_{CT1} = 1.5\ \Omega$ $R_{CT2} = 2\ \Omega$

The relay stability limit is normally, $16\ I_n$ then $I_{\text{through fault}}\ (I_F) = 16\ I_n$:

$$I_n = \frac{500 \times 10^3}{\sqrt{3} \times 500} = 577.35\ \text{A}$$

$$I_{F\ \text{through}} = 16 \times 577.35 = 9237.6\ \text{A}$$

Assuming one CT is saturated then we calculate the stability voltage of the relay (R_V) as follows:

$$V_S = \text{stability voltage} = \frac{I_{\text{through fault}}}{\text{CT}} \times \text{loop resistance}$$

$$\text{Loop}_1\ \text{Resistance} \quad = R_1 + R_2 + R_{CT1}$$
$$= 2 + 2 + 1.5 = 5.5\ \Omega$$

$$\text{Loop}_2\ \text{Resistance} \quad = R_3 + R_4 + R_{CT2}$$
$$= 3 + 3 + 2 = 8\ \Omega$$

$I_{\text{through fault}}$ 500 kV side $= 9237.6$ A.

$$I_{\text{through fault}}220\text{kV side} = 9237.6\frac{500}{220} = 20994.5\text{A}$$

$$\begin{aligned} V_{S1} &= \frac{I_{\text{through fault}}\ 500\ \text{kV side}}{\text{CT HV side}}.R_{\text{Loop1}} \\ &= \frac{9237.6}{600} = 5.5 = 84.678\ \text{V} \end{aligned}$$

$$V_{S2} = \frac{I_{\text{through fault}}220\ \text{kV side}}{\text{CT LV side}} = \frac{20,994.5}{1000}\ 8 = 167.956\ \text{V}$$

$V_{S2} > V_{S1}$

Then choose $V_S = V_{S2}$.

Use the relay with set voltage 25−175 V with 25-V taps. Choose $V_S = 170$ V:

$R_{V\ \text{burden}} = 1$ VA

Also choose CT knee point at least $2V_S$

$$\begin{aligned} VK_{\min} &= 2 \times 170\ \text{V} \\ &= 340\ \text{V} \end{aligned}$$

Choose $VK = 400$ V for CT, when another relay (current relay) R_1 is used with the voltage relay in parallel and given that the burden of current relay is 600 Ω and its current is 40 mA.

Then the voltage of the current relay (R_1) is $600 \times 0.040 = 24$ V.

Then the series setting resistor of the current relay R_1 can be calculated as follows (as shown in Fig. 19.5.8):

$$(R_S)\text{Series resistor of}R_I = \frac{(170 - 24)}{0.04} = 3650\Omega$$

The relay circuit consists of the following relays:

R_V = voltage-operated relay;
R_I = current-operated relay;
R_S = series setting resistor of R_I;
R_{shunt} = shunt resistor.

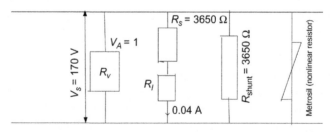

FIGURE 19.5.8   High-impedance differential relay circuit.

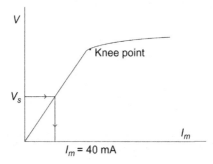

FIGURE 19.5.9   Current transformer magnetizing curve.

Metrosil resistance (nonlinear resistor)

We then check whether we need to use a shunt resistor for the relay or if this is not required as follows.

From the CT magnitude curve the value of $I_{\text{magnitising}}$ (I_m) can be checked at a setting voltage V_S for R_V at $V_S = 175$ V. I_m will be 40 MA, as shown in Fig. 19.5.9.

The total operating current of the relay = $I_{\text{CT HV side mag}} + I_{\text{CT low voltage side mag}} + I_{Rv} + I_{RI}$

$$I_{RV} = \frac{VA}{V} = \frac{1}{170} = 5.88 \text{ mA}$$

$I_{RI} = 40$ mA
$I_{\text{CT HV}} = 40$ mA, $I_{\text{CT LV}} = 40$ mA
Total operating current = $40 + 40 + 5.88 + 40 = 125.88$ mA

$$I_{\text{set}} = 0.2 \, I_{N \text{ HV side}}$$
$$= 0.2 \times \frac{500 \times 10^3}{\sqrt{3} \times 500} \times \frac{1}{600} = 0.192 = 192.45 \text{ mA}$$

Then a shunt resistor is required to increase the total operating current of the relay to 192.45 mA:

Current of shunt resistor = $192.45 - 125.88 = 66.57$ mA.

$$\text{Shunt resistor} = \frac{170V}{0.0666} = 2552.55 \Omega$$

Total operating current for setting value = $I_{\text{CT HV mag}} + I_{\text{CT LV mag}} + I_{RV} + I_{RI} + I_{R \text{ shunt}}$
$$= 40 + 40 + 5.88 + 40 + 66.57$$
$$= 192.45 \text{ mA}$$

Based on the available Metrosil resistance in the market, the following equation is used:

$$V_{\text{metrosil}} = (I_{\text{metrosil}})^{\beta} . C$$

where $\beta = 0.2$ $C = 450$

$I = 30$ A

$V_{\text{metrosil}} = (30)^{0/2} . 450 = 888$ V
So, for CT with 1 A secondary use metrosil with
$C = 450$, $\beta = 0.2$
I_{metrosil} will be 30 A for 2 seconds
$V_{\text{metrosil}} = 888$ V

19.5.3.2 Worked Example 2 (Fig. 19.5.10)

Autotransformer 500 MVA 500/220 kV
HV side CT$_1$ 600/1 $R_{\text{CT1}} = 1.5$ Ω
LV side CT$_2$ 1600/1 $R_{\text{CT2}} = 2.5$ Ω
Neutral CT$_3$ 1000/1 $R_{\text{CT3}} = 2.5$ Ω
Relay stability limit $= 16$ $I_n = I_{\text{through fault}}$

$$V_s = \text{stability voltage} = \frac{I_{\text{through fault}}}{\text{CT}} \times \text{loop resistance}$$

$$
\begin{aligned}
V_{S1 \text{ HV side}} &= 16 \times \frac{500 \times 10^3}{\sqrt{3} \times 500} (R_1 + R_2 + R_{\text{CT1}}) \\
&= \frac{16}{600} \times \frac{500 \times 10^3}{\sqrt{3} \times 500} (2 + 2 + 1.5) \\
&= 15.396 (5.5) = 84.678 \text{ V}
\end{aligned}
$$

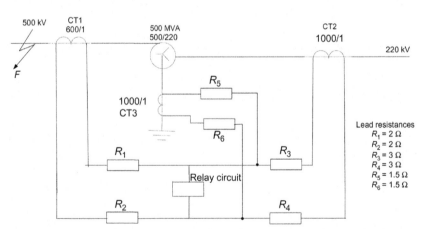

FIGURE 19.5.10 500/220 kV autotransformer 500 MVA protected by high-impedance differential protection.

$$V_{S2 \text{ LV side}} = \frac{16 \times 500 \times 10^3}{1000 \times \sqrt{3} \times 220} (R_3 + R_4 + R_{CT2})$$

$$= 20.994 \ (3 + 3 + 2.5)$$
$$= 20.994 \ (8.5) = 178.449 \text{ V}$$

For an autotransformer:

$$I_{\text{Neutral}} = I_{\text{LV side}} - I_{\text{HV side}}$$

$$V_{S \text{ neutral side}} = (20.994 - 15.396)(R_5 + R_6 + R_{CT3})$$
$$= 5.598 \ (1.5 + 1.5 + 2.5)$$
$$= 5.598(5.5)$$
$$= 30.789 \text{ V}$$

Then, $V_{S2} > V_{S1} > V_{S3}$.

Choose V_{s2} as $V_S = 180$ V.

Use voltage relay 20−200 V with 20-V taps.

Set $V_S = 180$ V, burden = 1 VA.

Then $V_{K \text{ min}}$ should be $2V_S$.

$V_{K \text{ min}} = 2 \times 180 = 360$ V.

Use CT with $V_K = 400$ V ,a current relay with burden = 500 Ω, and current = 30 mA.

The voltage across the current relay R_I will be as follows:

$V = 500 \times 0.03 = 15$ V

Then a series resistor is required to adjust the setting of the current relay R_I calculated as follows (as shown in Fig. 19.5.11):

$$R_S(\text{setting series resistor}) = \frac{180 - 15}{0.03}$$

$$R_S = 5500 \ \Omega$$

Check if shunt resistance is needed or not.

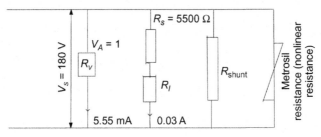

FIGURE 19.5.11 Relay circuit.

$$I_{\text{set}} = 0.2\,I_N \text{ HV side}$$
$$= 0.2 \times \frac{500 \times 10^3}{\sqrt{3} \times 500} \times \frac{1}{600}$$
$$= 0.1924 \text{ A}$$
$$= 192.24 \text{ mA}$$

From the CT magnitude curve check the value of I_m at $V_S = 180$ V $\rightarrow I_{\text{mag}} = 45$ mA.

The total operating current of the relay $= I_{Rv} + I_{RI} + I_{CT \text{ HV side mag}} + I_{CT \text{ LV side mag}} + I_R \text{ CT neutral mag}$ and total operating current $= I_{\text{total}}$.

$$I_{RV} = \frac{VA}{V} = \frac{1}{180} = 5.55 \text{ mA}$$

$$I_{RI} = 30 \text{ mA}$$

$$I_{\text{total}} = 5.55 + 30 + 45 + 45 + 45 = 170.55 \text{ mA}$$

Then we need a shunt resistor calculated as follows:
I of R shunt $= 192.24 - 170.55 = 21.69$ mA

$$R_{\text{shunt}} = \frac{180}{0.02169} = 8219.17 \ \Omega$$

$$I_{\text{total}} = I_{RV} + I_{RI} + I_{CT \text{ HV mag}} + I_{CT \text{ LV mag}} + I_{CT \text{ neutral mag}} + I_R \text{ shunt}$$
$$= 5.55 + 30 + 45 + 45 + 45 + 21.69$$
$$= 192.24 \text{ mA}$$

Subchapter 19.6

Generator Protection Setting

19.6.1 INTRODUCTION

As generators are a very costly component of the power system they should have a good protection system. The complexity of this protection system varies depending on the size of the units. The settings of these protections are set at different values depending on the utilities setting polices. Here we provide some examples of generator protection settings.

19.6.2 WORKED EXAMPLES

Worked Example 1 (as shown in Fig. 19.6.1)
Power station data:
Generator unit (G)

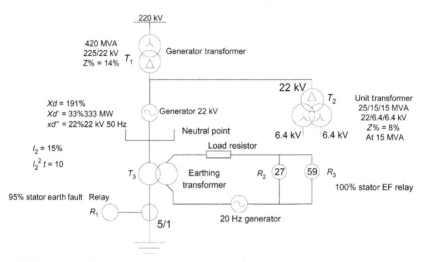

FIGURE 19.6.1 Generator and transformer data.

333 MW $xd = 191\%$ $xd' = 33\%$
$Xd'' = 22\%$ F = 50 Hz
$I_2 = 15\%$ $I^2.t = 10$ S CT = 13,000/1−22 kV
Power Factor $(PF) = 0.8$ S = 417 MVA
Unit transformer (T_2)
25/15/15 MVA 22/6.4/6.4 kV
$Z\% = 8\%$ at 15 MVA
Generator transformer (T_1)
420 MVA 225/22 kV $Z\% = 14\%$
Station transformer (T_3)
25/15/15 MVA
220/6.4/6.4 kV
$Z\% = 8.2\%$ at 15 MVA
CTs = 13,000/1
1. Stator earth fault protection
95% stator earth fault relay (R_1)
Maximum fault current in neutral = 20 A
$I_{set} = (1-0.95) \, 20.\frac{1}{5} = 0.2$ A
t to be set to = 0.05 second.
Tripping action:

1. HV CB of transformer T_1;
2. T_2 6.4 kV incoming feeder tripping;
3. Turbine tripping;
4. Generator excitation tripping.

2. 100% stator earth fault protection
R_2 and R_3 will be two relays as follows:

1. Undervoltage relay -27 works on third harmonic (R_2).
2. Overvoltage relay (R_3) works on fundamental frequency.

The overlap between 27 and 59 covers 100% of the stator winding.
3. Stator OC protection
$I_{set} = 1.05\ I_N\ P = 333$ MW then $S = 417$ MVA at a power factor of 0.8.

$$I_N = \frac{417 \times 10^3}{\sqrt{3} \times 22} = 10,943\ \text{A}$$

$$I_{N\ \text{secondary}} = 10,943 \times \frac{1}{13,000} = 0.842\ \text{A}$$

$$
\begin{aligned}
I_{set} \quad & 1.05\ I_N \\
& 1.05 \times 0.842 \\
& = 0.884\ \text{A}
\end{aligned}
$$

$$
\begin{aligned}
\text{Set} I_{set} \quad & = 0.9\ \text{A} \\
t & = 6\ \text{seconds}
\end{aligned}
$$

Action: trip of CB of transformer (T_1) 220 kV side.
4. Negative-phase sequence current I_2 protection (46)
Second stage (trip):

$$
\begin{aligned}
I_{set} \quad & = 0.15\ I_N \\
& = 0.15 \times 0.842 = 0.1263\ \text{A} \\
I_{set} \quad & = 0.127\ \text{A} \\
\text{Set}\ I_{set} \quad & = 12\%
\end{aligned}
$$

$I^2_2.t = 10$ where $K = I^2_2.t$, K factor $= 10$.
First stage (alarm):

$$
\begin{aligned}
\text{Set}\ I_{alarm} = 0.9\ I_{set} \quad & = 0.9 \times 0.12 \\
& = 0.108 \\
& = 10.8\%
\end{aligned}
$$

Set $t_{alarm} = 3.0$ seconds
Action: trip 220 kV side CB of T_1 second stage.
Alarm in first stage.
5. Loss of excitation protection
Loss of excitation leads to loss of synchronism and running at higher than synchronous speed.
The following impedance relay is used (Fig. 19.6.2).
$X_1 + x_2 = xd$
$X_1 = xd - x_2$
$xd' = 33\%$ on 417 MVA base.

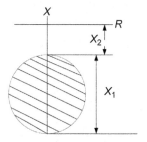

FIGURE 19.6.2 Loss of excitation impedance protection for generators.

$$X_2 = \frac{xd'}{2}$$

$Xd = 191\%$ on 417 MVAb.

$$Z_{(\Omega)} = \frac{Z\% \text{ MVAb}.100}{(VL)^2}$$

$$Xd' \ (\Omega) = \frac{33 \times 417.100}{100(22)^2} = 28.43 \ \Omega$$

$$X_2 = \frac{xd'}{2} = \frac{28.43}{2} = 14.21 \ \Omega$$

$X_1 = xd - x_2$

$$Xd \ (\Omega) = \frac{191.417 \times 100}{100.(22)^2} = 164 \ \Omega$$

$X_1 = 164 - 14.21 = 150.35 \ \Omega$
$X_1 = 150.35 \ \Omega$ primary, secondary $x_1 = 887.065 \ \Omega$
$x_2 = 14.21 \ \Omega$ primary, secondary $x_2 = 83.839 \ \Omega$

$$t = 1 \text{ second}, \ Z_s = Z_P \frac{NV}{NC} = 5.9 \ Z_P$$

$$\text{VT ratio} = \frac{220,000/\sqrt{3}}{100/\sqrt{3}} = 2200 = NV$$

CT ratio $= 13,000/1 = NC$

6. Underimpedance protection

This relay is used as a backup protection for external faults on the high-voltage side (220 kV) of a transformer (T_1), as shown in Fig. 19.6.3.

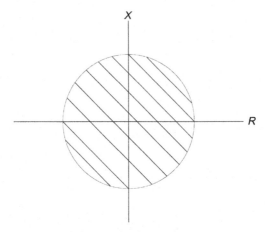

FIGURE 19.6.3 Backup impedance protection for generators.

$$Z_{set} = 0.7 \times 0.36 = 0.252 \rightarrow = 25.2\%$$

$$Z_{set} (\Omega) = \frac{25.2 \times 417 \times 100}{100(220)^2}$$
$$= 0.217 \ \Omega \ (\text{primary})$$

$$Z_{set}(\Omega) \ \text{secondary} = 5.9 \ Z_{primary}$$
$$= 5.9 \times 0.217$$
$$Z_{set} = 1.2803 \ \Omega$$

$$Z_{set} = 0.7 \ Z_{Load}$$
$$Z_{Load \ min} = XG + XT$$
$$= Xd'' + XT$$
$$= 0.22 + 0.14$$
$$= 0.36$$

Set time = 3 seconds.
Action:

1. Turbine tripping;
2. Tripping of 6.4 K unit transformer incoming;
3. Generator excitation tripping.

7. Generator frequency protection
A. Underfrequency protection
This protection is required for load-shedding to allow unit frequency recovery.

Relay set = 47 Hz F = 50 Hz
t_{set} = 4 seconds
Action:

1. Trip 220 kV CB. This protection works when load-shedding of the network does not work.

8.Overfrequency protection
Under this condition the unit overspeed and trip are set as follows.
Relay set = 54 Hz, F = 50 Hz
Action:

1. Trip of 220 kV CB.

8. Undervoltage stator protection
This protection works as a backup for an HV network in two stages.
Stage 1 has a definite time uv set to 0.7 V_n of the generator and trips the HV CB.
Stage 2 trips the unit after 2.5 seconds.
Stage 1:

$$V_{set} = 0.7V_n = 0.7 \times 100 = 70 \text{ V}$$

t_{set} = 1.2 second
Action trip CB of 220 kV transformer (T_1).
Stage 2:
V_{set} = 70 V
t_{set} = 2.5 seconds
Action:

1. Turbine tripping;
2. Tripping of 6.4 kV transformer incoming;
3. Generator excitation tripping.

9. Overvoltage protection
This protection works when the Automatic Voltage Regulator (AVR) maloperates or for manual operation of the excitation system.
First stage

$$V_{set} = 1.1 \ V_n = 1.1 \times 100 = 110 \text{ V}$$

t = 4.5 seconds
Action: alarm
Second stage

$$V_{set} = 1.2 \ V_n = 120 \text{ V}$$

t_{set} = 3 seconds

Action:

1. HV CB tripping of T_1;
2. Incoming of transformer T_2 6.4 kV trip;
3. Turbine tripping;
4. Generator excitation tripping.

10. Generator differential protection
A biased differential relay is used in the following circumstances
(Fig. 19.6.4)

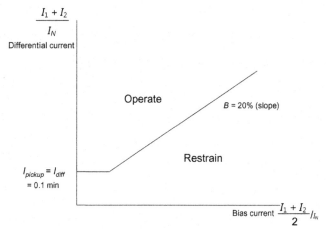

FIGURE 19.6.4 Biased differential protection for generators.

(I_N = generator nominal current):
$$P = \sqrt{3}\ V_L.\ I_N.\ \cos \varnothing$$

$$I_N = \frac{P}{\sqrt{3}V_L.\cos \phi} = \frac{333 \times 10^3}{\sqrt{3}V_L.\cos \phi}$$

$$= \frac{333 \times 10^3}{\sqrt{3} \times 22 \times 0.8} = 10,924\ \text{A}$$

$$I_{N\ \text{secondary}} = \frac{10,924}{13,000} = 0.84\ \text{A}$$

$$I_{\text{set}} = 0.1\ I_N$$
$$= 0.1 \times 0.84$$
$$= 0.084\ \text{A}$$

Set bias stop = 20%
11. Reverse power protection
$$P_{\text{Reverse set}} = 0.01 \times 333 = 3.33\ \text{MW}$$

$$I_{\text{set on Relay}} = \frac{P_{\text{set}}}{\sqrt{3}V_L.\cos\phi}$$

$$= \frac{3.33 \times 10^3}{\sqrt{3} \times 22 \times 0.8}$$

$$= 109.237\,\text{A}$$

$$I_{\text{set secondary}} = \frac{109.237}{13000}$$

$$= 8.4\,\text{mA}$$

$t_{\text{set}} = 15$ seconds
Action:

1. Trip HV CB 220 kV of T_1;
2. Generator excitation trip;
3. Incoming 6.6 kV of T_2 trip.

Subchapter 19.7

Switchgear (Busbar) Protection Settings

19.7.1 INTRODUCTION

The busbars in substations are a very important part of the system and should have adequate protection, however this protection is very expensive in 11 kV and 22 kV, and the faults in busbars are rare, therefore it is used in 66 kV systems and above as essential protection.

The most commonly used scheme for busbar protection is high-impedance differential protection.

19.7.2 WORKED EXAMPLES FOR BUSBAR PROTECTION

One hundred and thirty-two kiloVolts busbars have two sections and a bus coupler CB, each busbar has one feeder and an incomer transformer (as shown in Fig. 19.7.1).

CT ratios:
CT_1, CT_2, CT_3, CT_4, CT_7, CT_8, CT_9, $CT_{10} \rightarrow 1200/1$ $R_{CT} = 2.9\,\Omega$
$I_{\text{mag}} = 40$ mA at $V = 130$ V $I_{\text{mag}} = 35$ mA at $V = 125$ V CT ratios:
CT_5, $CT_6 \rightarrow 1500/1$
$R_{CT} = 3.1\,\Omega$
$I_{\text{mag}} = 45$ mA at $V = 130$ V
Lead resistances:
$R_1 = 0.3\,\Omega$ $R_4 = 0.32\,\Omega$ $R_7 = 0.33\,\Omega$

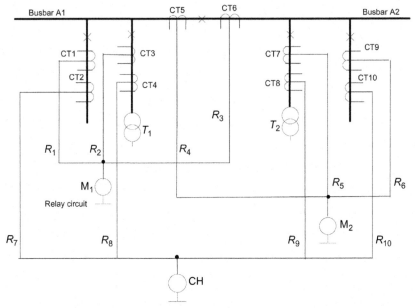

FIGURE 19.7.1 Two-section busbar with two feeders and two incomers.

$R_2 = 0.35 \, \Omega$ $R_5 = 0.36 \, \Omega$ $R_8 = 0.31 \, \Omega$
$R_3 = 0.4 \, \Omega$ $R_6 = 0.48 \, \Omega$ $R_9 = 0.33 \, \Omega$ $R_{10} = 0.39 \, \Omega$
We use three relays as follows:

(M_1) Discriminating main 1 relay to protect busbar section A1;
(M_2) Discriminating main 2 relay to protect busbar section A2.

Check zone relay (CH) to protect busbar A1 and A2 to confirm that the fault exists on the busbars.

Relay setting voltage $= \frac{I_F}{CT} R_{\text{loop max}}$
$R_{\text{loop max}} =$ maximum lead resistance
Assuming saturation in one CT:
M_1 (main relay setting)

$$
\begin{aligned}
R_{\text{lead 1}} &= 2 \times R_1 + R_{\text{CT1}} \\
&= 2 \times 0.3 + 2.9 = 3.5 \, \Omega
\end{aligned}
$$

$$
\begin{aligned}
R_{\text{lead 2}} &= 2 \times R_2 + R_{\text{CT3}} \\
&= 2 \times 0.35 + 2.9 = 3.6 \, \Omega
\end{aligned}
$$

$$
\begin{aligned}
R_{\text{lead 3}} &= 2 \times R_3 + R_{\text{CT6}} \\
&= 2 \times 0.4 + 3.1 = 3.9 \, \Omega
\end{aligned}
$$

$R_{\text{lead max}} = R_{\text{lead 3}} = 3.9 \, \Omega$
Maximum through fault current $= 40$ kA

$$V_S = \frac{I_F}{\text{CT}} \cdot R_{\text{loop max}}$$

$$= \frac{40,000}{1200} \, 3.9 = 130 \text{ V}$$

Choose voltage relay R_V as follows:

Range $25-175$ V $I_{RV} = 35$ mA

Set $V_S = 130$ V

Then the knee point voltage of CT should be as follows:

$$V_{K \text{ min}} = 2 \times V_S$$
$$= 2 \times 130 = 260 \text{ V}$$

Choose $V_K = 300$ V

Use a current relay with burden $= 3500 \, \Omega$

$I = 30$ mA

Then the voltage across the current relay R_I is as follows:
$3500 \times 0.03 = 105$ V.

Then, the series resistor is required to adjust the setting of the current relay R_I and is calculated as follows (as shown in Fig. 19.7.2):

$$R_S(\text{Setting Resistor}) = \frac{130 - 105}{0.03} = 833 \, \Omega$$

M_1 relay circuit—busbar protection

We then need to check whether we need shunt resistance or not:

$I_{\text{set}} = 0.15 \, I_{\text{fault min}}$

Assume the single-phase minimum fault $= 12$ kA, and the three-phase minimum fault $= 14$ kA:

$$I_{\text{set}} = 0.15 \times 12 = 1.8 \text{ kA}$$

$$I_{\text{set secondary}} = \frac{1800}{1200} = 1.5 \text{ A}$$

$I_{\text{total operating current}} = I_{RV} + I_{RI} + I_{\text{CT1 mag}} + I_{\text{CT3 mag}} + I_{\text{CT6 mag}}$

$I_{RV} = 35$ mA $I_{RI} = 30$ mA

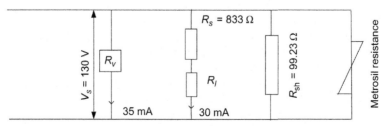

FIGURE 19.7.2 Relay circuit.

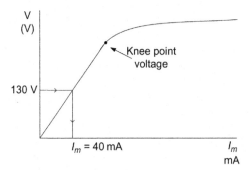

FIGURE 19.7.3 Current transformer magnetization curve.

By checking from the magnetization curve of CT, the magnetization current at $V_S = 130$ V (Fig. 19.7.3).

$I_{total} = 35 + 30 + 40 + 40 + 45 + I_{R\ shunt} = 190$ mA $+ I_{R\ shunt}$

$I_{total} = I_{set} = 1.5$ A $= 1500$ mA

$I_{R\ shunt} = 1500 - 190 = 1310$ mA

$$R_{shunt} = \frac{V_S}{I_{R\ shunt}} = \frac{130}{1310/1000} = 99.23\ \Omega$$

$V_S = 130$ V

Choose Metrosil as follows:

$V_{metrosil} = (I_{metrosil})^{\beta}\ C$

Choose $C = 450$ $\beta = 0.25$ $I = 30$ A

$$V_{metrosil} = (30)^{0.25} \times 450$$
$$= 1053\ V$$

$I_{metrosil}$ will be 30 A for 2 seconds.

Main 2 relay setting M_2

$$R_{lead1} = 2 \times R_6 + R_{CT9}$$
$$= 2 \times 0.48 + 2.9$$
$$= 3.86\ \Omega$$

$$R_{lead\ 2} = 2 \times R_5 + R_{CT7}$$
$$= 2 \times 0.36 + 2.9 = 3.62\ \Omega$$

$$R_{lead3} = 2 \times R_4 + R_{CT5}$$
$$= 2 \times 0.32 + 3.1 = 3.74\ \Omega$$

$R_{lead\ max} = R_{lead\ 1} = 3.86$

$$V_S = \quad = \frac{I_F}{CT} R_{\text{loop max}}$$

$$= \frac{40,000}{1200} \, 3.86 = 128.66 \text{ V}$$

$V_S = 130$ V
Choose relay $R_V = 130$ V, $I_{RV} = 35$ mA
Range: $25-175$ V
Knee point of CT is as follows:

$$V_{K \min} = 2V_S = 2 \times 130 = 260 \text{ V}$$

Choose CT with $V_k = 300$ V
Use a current relay with burden 3500 Ω, $I = 30$ mA.
As before in the M_1 calculation:
$R_{\text{sh}} = 99.23$ Ω $V_{\text{metrosil}} = 1053$ V
$R_{\text{series Resistor}} (RS) = 833$ Ω $C = 450$, $\beta = 0.25$
$I_{\text{metrosil}} = 30$A for 2 seconds
Then check the zone relay CH calculation:

$$R_{\text{lead 1}} \quad = 2 \times R_7 + R_{CT2}$$
$$0.33 \times 2 + 2.9 = 3.56 \ \Omega$$

$$R_{\text{lead 2}} \quad = 2 \times R_8 + R_{CT4}$$
$$2 \times 0.31 + 2.9 = 3.52 \ \Omega$$

$$R_{\text{lead 3}} \quad = 2 \times R_9 + R_{CT8}$$
$$= 2 \times 0.33 + 2.9 = 3.56 \ \Omega$$

$$R_{\text{lead 4}} \quad = 2 \times R_{10} + R_{CT10}$$
$$= 2 \times 0.39 + 2.9 = 3.68$$

$R_{\text{lead max}} = R_{\text{lead 4}} = 3.68$

$$V_S \quad = \frac{I_F}{CT} R_{L \max}$$

$$= \frac{40,000}{1200} \, 3.68$$

$$= 122.66$$

Set $V_S = 125$ V
Choose the relay range: $25-175$ V
$V_{\text{set}} = 125$ V $I_{RV} = 35$ mA
The knee point voltage of CT should be as follows:

$$V_{K \min} = 2 \times V_S = 2 \times 125 = 250 \text{ V}$$

Choose $V_K = 300$ V of CT using a current relay with burden $= 3500$ Ω

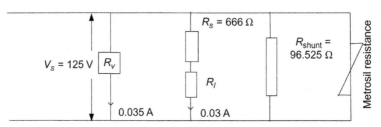

FIGURE 19.7.4 Check the zone (CH) relay circuit.

$I = 30$ mA

Then, the voltage across the current relay R_I will be as follows:

$$= 3500 \times 0.03 = 105V$$

The series setting resistor R_S is calculated as follows (as shown in Fig. 19.7.4):

$$R_S = \frac{125 - 105}{0.03} = 666 \ \Omega$$

Then check whether we need the shunt resistance or not:

$$I_{set} = 0.15 \ I_{fault \ min}$$
$$= 0.15 \times 12 = 1.8 \ KA$$

$$I_{set \ secondary} = \frac{1.8 \times 10^3}{1200} = 1.5 \ A$$

$I_{total \ operation} = I_{RV} + I_{RI} + I_{CT2 \ mag} + I_{CT4 \ mag} + I_{CT8 \ mag} + I_{CT10 \ mag} + I_{R \ shunt}$
$I_{total} = 35 + 30 + 35 + 35 + 35 + 35 = 205$ mA
$I_{mag} = 40$ mA for CT_2, CT_4, CT_8, CT_{10} from $V-I_{mag}$. The curve of CT at $V_S = 125$ V
$I_{mag} = 35$ mA
$I_{total} = 205 + I_{R \ shunt}$

$$I_{total} = I_{set} = 1.5 \times 10^3 = 1500 \ mA$$

$I_{R \ shunt} = 1500 - 205 = 1295$ mA

$$R_{shunt} = \frac{VS}{IRshunt} = \frac{125}{1295/1000}$$

$R_{shunt} = 96.525 \ \Omega$
Choose metrosil resistance with voltage:
$V_{metrosil} = (I_{metrosil})^\beta \ C$
$C = 450$, $\beta = 0.25$, $I = 30$ A
$V_{metrosil} = (30)^{0.25} \ 450 = 1053$ V
Relay setting

Main 1 zone (discriminating zone)—M$_1$
R_V V_{range} 25–175 V, I_{RV} = 35 mA, V_{set} = 130 V
R_I burden = 3500 Ω, I_{RV} = 30 mA, R_S = 833 Ω
R_{shunt} = 99.23 Ω, Metrosil C = 450 β = 0.25 I = 30 A at 2 seconds.
Main 2 zone (discriminating zone)—M$_2$
R_V V_{range} = 25–175 V, I_{RV} = 35 mA, V_{set} = 130 V
R_I burden = 3500 Ω, I_{RV} = 30 mA, R_S = 833 Ω
R_{shunt} = 99.23 Ω, Metrosil C = 450, β = 0.25, I = 30 A at 2 seconds.
Check zone (CH)
R_V V_{range} = 25–175 V, I_{RV} = 35 mA, V_{set} = 125 V
R_I burden = 3500 Ω, I_{RV} = 30 mA, R_S = 666 Ω
R_{shunt} = 96.525 Ω, Metrosil C = 450, β = 0.25, I = 30 A at 2 seconds.

19.7.3 INTRODUCTION TO BREAKER FAILURE PROTECTION

This protection is used as backup protection to isolate a fault if the circuit breaker of the feeder has a mechanical or other problem preventing it from opening the circuit. This protection then isolates all sources which feed the fault and also the remote end of the line.

19.7.4 WORKED EXAMPLE FOR BREAKER FAILURE PROTECTION

CB failure time
    t_{CB} = 110–180 ms
    Fault clearance time should not exceed 300 ms
    Worked example (Figs. 19.7.5 and 19.7.6):
    Breaker Failure (BF) relay trip logic
    BF relay operating times
    Refer to Fig. 19.7.7 for the breaker failure time and the total fault clearance time.
    $I_{set\ BF}$ = 0.5 I_N

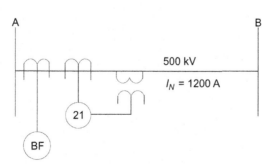

FIGURE 19.7.5 Breaker failure relay connection on an overhead transmission line.

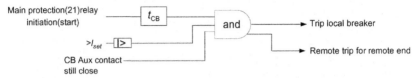

FIGURE 19.7.6 Breaker failure relay operation logic.

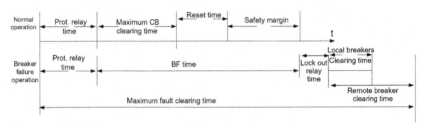

FIGURE 19.7.7 Breaker failure time and total fault clearance time.

$= 0.5 \times 1200 = 600$ A

$t_{CB} = 120$ ms

Action:

The back trip for all feeders connected on busbar A through busbar protection trip relays for the section which this feeder connected to it should not exceed 300 ms.

The remote trip for the breaker is at end B.

The local retrip for the breaker is at end A.

Subchapter 19.8

Motor Protection Setting

19.8.1 INTRODUCTION

Here we provide some examples for motor protection settings.

19.8.2 WORKED EXAMPLES

19.8.2.1 Motor Faults

1. Winding faults
2. Overloading;
3. Reduced or loss of supply voltage;
4. Phase reversal;

5. Phase unbalance;
6. Out-of-step operation for synchronous motors;
7. Loss of excitation for synchronous motors.

19.8.2.2 Phase Fault Protection

A high set Over Current (OC) unit is used.

Differential protection is also used as it is not affected by the motor starting current.

For the H set OC relay the setting should be more than the motor starting current:

$I_{set} > I_{start}$.

19.8.2.3 Earth Fault Protection

An inverse or very inverse OC relay is connected to the residual CT circuit (as shown in Fig. 19.8.1).

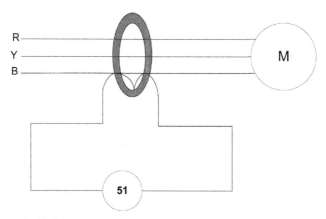

FIGURE 19.8.1 Earth fault protection for motors.

$I_{set} = 0.25\, I_{F\,min}$, where $I_{F\,min}$ = minimum fault current.

19.8.2.4 Locked Rotor Protection

We use the OC relay with characteristics higher than the starting current of the motor and less than the locked rotor current (as shown in Fig. 19.8.2), where t_s = motor starting time; t_R = locked rotor time.

Also, the OC (51) relay characteristic is used to detect the locked rotor condition of the motor or impedance relay is used to detect this condition, as shown below in Fig. 19.8.3.

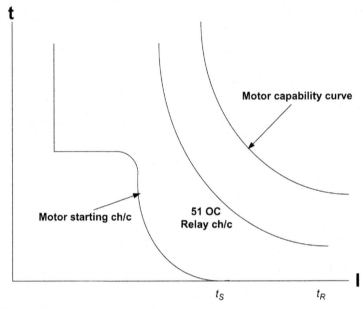

FIGURE 19.8.2 Locked rotor protection for motors.

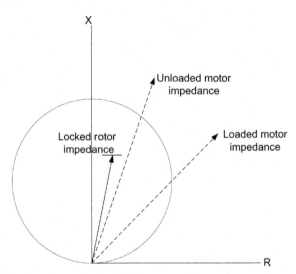

FIGURE 19.8.3 Impedance relay for locked rotor condition.

19.8.2.5   Overload Protection

Thermal replica relays are used to protect the motor from overloads using a separate relay in each phase of the motor. Thermal relays are also used with thermal replica relays (resistance temperature detectors embedded in machine winding) to give complete protection for the motor.

19.8.2.6   Low-Voltage Protection

Undervoltage time delay relay is used as follows:
$V_{set} = 0.5$ V nominal
Action: Trip after a few seconds.

19.8.2.7   Phase Rotation Protection

A negative-phase sequence voltage relay is used.
$v_{2\ set} = 10\%\ V_2$

19.8.2.8   Phase Unbalance Protection

Detect I_2 and set as follows:
$I_{set} = 0.15\ I_N$
To protect the motor against single-phasing condition:
$I_2^2\ t = 40$ in general

19.8.2.9   Out-of-Step Protection for Synchronous Motors

A relay is used to detect the condition in which the motor draws a heavy current in a low-power factor. Therefore, this relay will be a power factor relay and will issue a signal to disconnect the field winding and let the motor run as an induction motor.

19.8.2.10   Loss of Excitation Protection for Synchronous Motors

Undercurrent relay is inserted in a field circuit.

Chapter 20

Protective Relays Testing and Commissioning

INTRODUCTION

In this chapter we explain the required testing at startup of a plant or substation for protective relays.

Subchapter 20.1

Protective Relays Testing and Commissioning

20.1.1 INTRODUCTION

There are different types of relay testing as described below.

The several standard tests done in a factory are called type tests and to be done when we introduce a relay in a utility network.

Commissioning or site acceptance test are less in their steps than the above type tests and more related to the actual operation of relays at the network. *Routine maintenance tests* are done internally in regular time to detect any faults in the protective relays. Electromechanical relays require these tests in more regularly as these relays have more moving parts than static digital relays which have self testing facilities which can detect any fault on the relay components to give an alarm and block operation of the relay when the fault affects the proper operation of the relay.

20.1.2 COMMISSIONING TESTS OF PROTECTIVE RELAYS

In general these tests include the following.

Practical Power System and Protective Relays Commissioning.
DOI: https://doi.org/10.1016/B978-0-12-816858-5.00020-4

20.1.2.1 Secondary Injection Tests

These tests confirm the relay characteristic and check the relay pick-up accuracy.

20.1.2.2 Protection Scheme Function Test

This test confirms the operation of the relay with other relays, auxiliary relays, and other scheme components of protection. This test is done not only in the relay panel itself but is also done between the relay panel and other relays panels, for example, the busbar protection panel and the feeder main-1 protection panels.

20.1.2.3 Primary Injection Tests

In these tests the complete system of protection is tested. The operation of current transformer (CT), voltage transformer (VT) and protection relays also will be tested, for example, checking the stability of the transformer differential and restricted earth fault protection by injection 380 V and also busbar protection sensitivity and stability tests. These tests are explained in the following chapters of this book in detail.

20.1.2.4 End-to-End Test

To be done on transmission lines differential protection and distance relay telecommunications tripping schemes. These tests need two engineers at each end of the line to simulate normal and fault conditions and to check the relay response at both ends of the line and also to check the decision for tripping the circuit breakers (CBs) at the two ends of the line.

20.1.2.5 On-Load Tests

These tests are performed after energization of a new circuit and can include a directionality test to check that the relay (directional over current (OC) relays or distance relays) works in the line direction and not in the busbar direction, whilst also checking the operation of line differential relays on the load. These load tests are done after opening the tripping links of the relay but keeping the circuit protected by another protection on the same circuit which the relay under test protects. These tests are explained in the following chapters.

Subchapter 20.2

Overcurrent Relay Testing and Commissioning

20.2.1 SECONDARY INJECTION TESTS

20.2.1.1 Nondirectional Overcurrent Relays Testing

Test Connection

As shown in Fig. 20.2.1 to test an OC relay in service we need to isolate the links A—B and C—D and to short the CT secondary circuit by closing the link A—C.

As we know, during CT loading if there is an open circuit this will generate a dangerously high voltage (HV) on the secondary of the CT—dangerous to personnel and capable of damage to the CT insulation (sometimes exceeding 15—20 kV)—this exists in an AC relay circuit.

The injecting leads of the relay test set are then connected to the relay side. This means B&D points to R&N points on the test set, respectively, for the AC circuit of the relay and for the DC circuit of the relay we isolate the relay contact "trip contact" by opening the two isolating links in the relay

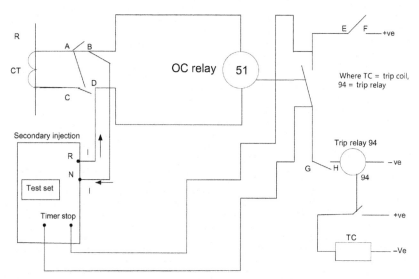

FIGURE 20.2.1   Single-phase pickup/drop out test for OC relay connections.

trip circuit E−F and G−H. Then we connect relay trip contact to the timer stop points of the tester as dry or No-Voltage contact.

1. Relay data:

   11 kV OC relay $I_{set} = 1.2 I_n$
   CT ratio = 1000/1 relay characteristic used is SI
   TD = 1 I_{sc} or $I_{inst} = 6 I_n$
   $I_{pickup} = I_{set} \pm 5\%$
   $I_{drop\ out} > I_{set} - 5\% I_{set}$
   where I_n = relay nominal current = 1 A

2. Pickup/drop off test:

   1. Inject a current in the secondary circuit of the relay and increase the current gradually from 0 up to 1.2 A. Then, when the relay led illuminates, put this value as the pickup value, say 1.26 A, then decrease the value of the injected current until the relay led switches off then record it as a drop value, say 1.14 A.
   2. Check that the value of I_{pickup} within the value of $I_{set} \pm 5\%$
      $I_{pickup} = 1.26 \rightarrow 1.2 + 0.05 \times 1.2 = 1.26$. Then it is correct.
   3. Check the drop out value $\geq I_{set} - 5\% I_{set}$

   $I_{drop\ out} = 1.14 A \rightarrow 1.2 - 0.05 \times 1.2 = 1.4$. Then it is correct.

   Check this for each phase of the relay. This means phases (Y) and (B). If we have a modern test set we can inject three phases at the same time and check the pickup and drop out values from the instantaneous contact of the relay. As shown in Fig 20.2.2 the test for pH−pH values is repeated.

3. Checking of the relay characteristic:

   1. Inject the current at two times the I_{set}, three times the I_{set}, and four times the I_{set}.
   2. Check the trip time at each value.
   3. From the relay characteristic curve check the value of times at PM = 1.0
      for $2 \times I_{set}$, $3 \times I_{set}$, $4 \times I_{set}$.
   4. Check that the difference between measured times and curve times is within $\pm 5\%$, repeat the test for pH−pH injection.
   5. Instantaneous element test:
      For this test inject the current and switch off very fast as this current setting is high (6 I_n) and can damage the relay circuit.
      Then inject 6 A and check the operating time in milliseconds.
   6. Fill in the test results as shown in the test sheet in Table 20.2.1.

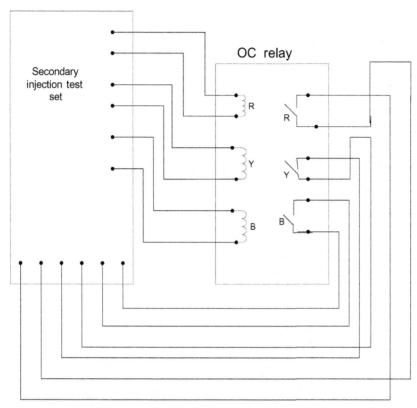

FIGURE 20.2.2 Three-phase pickup/drop out test connection for OC relay.

TABLE 20.2.1 Relay Test Sheet 1

Phase	R—N	Y—N	B—N	R—Y	Y—B	B—R
Pickup						
Drop out						
Multiplier		$2 \times I$		$3 \times I$		$4 \times I$
Measured time		-------		-------		-------
R—N						
Y—N						
B—N						
R—Y						
Y—B						
B—R						
Phase	R—N	Y—N	B—N	R—Y	Y—B	B—R
$I_{instantaneous}$ (A)						

20.2.1.2 Directional Overcurrent Relay Testing and Commissioning

Directional Over current (DOC) relay test connections are shown in Fig 20.2.3:

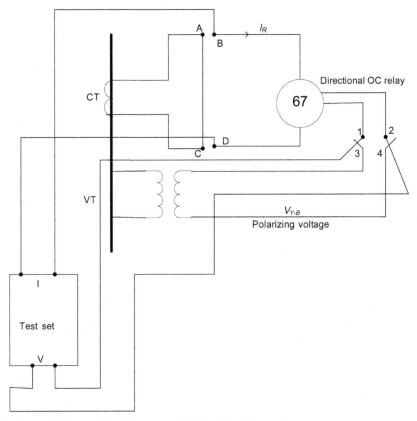

FIGURE 20.2.3 Directional overcurrent relay test circuit connections.

Close link A−C;
Open link A−B;
Open link C−D;
Open link 1−3;
Open link 2−4.

1. Repeat the same procedure to short circuit (SC) the current circuit for each phase and inject current at the relay side.
2. Open the VT secondary circuit in isolating link terminals and inject the voltage at the relay side.
3. Isolate the trip circuit by opening the links and connect the tester timer stop points to the trip contact of the relay.

1. Relay data:

11 kV OC relay $I_{set} = 1.2\ I_n$

CT ratio = 1000/1 relay characteristic used = SI

TD = 1 $I_{set} = 6\ I_n$

Relay characteristic angle = −45 degrees current logging voltage→ (90 degrees connection relay), as shown in Fig. 20.2.4.

Test the relay pickup/drop out and characteristic times and instantaneous current as per the tests done in Section 20.2.1.1 for nondirectional OC relay considering an injection of polarizing voltage and current with an angle in the operating zone.

For testing of the directional unit in phase (R) of the relay, inject a current greater than the pickup value, with a polarizing voltage from phase Y and B, then check the operating boundaries of the relay between lag 135 degrees and lead 45 degrees. Check that the error in angle is about ± 5 degrees, then fill in the test sheet in Table 20.2.2.

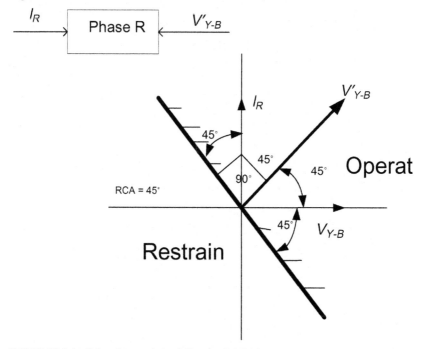

FIGURE 20.2.4 Relay characteristic of directional OC relay.

TABLE 20.2.2 Relay Test Sheet 2

Relay Characteristic Angle	−45 Degrees
Boundaries lag	
Boundaries lead	

Subchapter 20.3

Distance Protection Testing and Commissioning

20.3.1 SECONDARY INJECTION TEST

In this test we will inject voltage and current (fixed current) as per the manufacturer's test.

Recommendation 1 or 2 A. For the Current Transformer (CT) with rated current of 1 A we decrease the voltage until we reach the zone reach for pH-E and pH−pH characteristic for each phase R, Y, B and R-Y, Y-B, B-R phases as shown below:

1. Relay data

   CT = 2000/1 500 kV Transmission Line (TL)

   Phase-E characteristic Mho relay, pH−pH characteristic, quadrilateral characteristic.

   Z_1, Z_2, Z_3 are given by relay setting engineer for pH-E and pH−pH where:

   K_0 = residual earth compensation factor.

$$K_0 = 0.6 = \frac{(Z_0 - Z_L)}{3Z_L}$$

   Z_1 = positive-sequence impedance of TL
   Z_0 = zero-sequence impedance of TL

$$\text{VT ratio} = \frac{500,000}{\sqrt{3}} / \frac{100}{\sqrt{3}}$$

$$Z_s = Z_p \left(\frac{\text{CT ratio}}{\text{VT ratio}} \right) = Z_p \frac{2000}{5000}$$

   ($Z_s = Z_p$ 0.4) and Z_3 reverse = $0.25Z_1 = 5.25 \angle 260$ degrees
   $Z_1 = 21 \ \Omega \angle 80$ degrees $t_1 = 0$
   $Z_2 = 26 \ \Omega \angle 80$ degrees $t_2 = 500$ ms
   $Z_3 = 33 \ \Omega \angle 80$ degrees $t_3 = 1000$ ms
   Z power swing = $Z_{ps} = 1.3Z_3 = 42.9 \angle 80$ degrees blocking action after $t = 5$ seconds.
   Fig 20.3.1 Relay characteristic for pH-E and pH−pH.

2. Pretest procedures:
   1. The current circuit of the CT secondary should be short-circuited if the relay is to be tested in load condition.

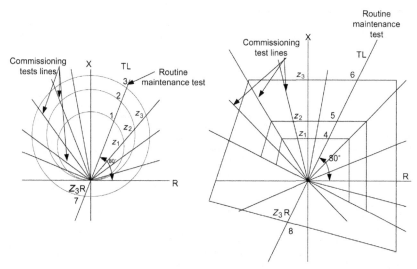

FIGURE 20.3.1   (A) Mho-relay characteristic for pH-E faults; (B) quadrilateral-relay characteristic for pH−pH faults.

2. We should disconnect VT secondary circuit from the relay circuit in the loading condition or switched off, as if the VT circuit secondary is left closed and we inject voltage in the relay, then if the VT is a capacitive type, a high current can be drown from the test set and circuit fuses can be blown for the tester.

3. Use a test switch if available to make the AC and DC connection for the relay and CT and VT isolation.

4. Isolate the tripping contacts of the relay and the start signal of the breaker failure relay before any starting any tests.

5. Isolate the autoreclosing scheme during the test.

6. Stop fault recorder initiation signals before the test to save unwanted operation, which consumes memory and paper.

7. Check by calculation the reach of each zone and compare with the relay setting sheet supplied by the relay setting engineer.

8. Put the values of the relay setting by hand into the electromechanical relay or by loading the setting file to the digital relay.

3. Relay connections are as shown in Fig. 20.3.2.

4. Zones reach test and tripping time

For routine maintenance test the relay should be tested in a line angle at points 1, 2, and 3 for pH-E and points 4, 5, and 6 for pH−pH faults.

However, for commissioning this test should be done in different line angles to check the 360 degrees distance relay zone reaches, and for the new advanced tester all points can be checked in a few hours, which previously took up to a few days.

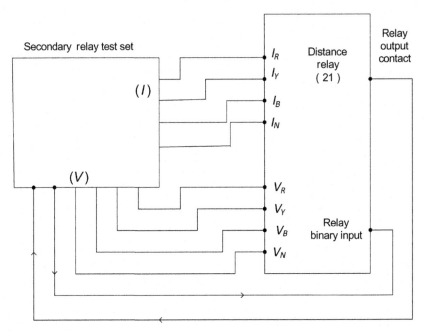

FIGURE 20.3.2   Relay test connections.

This test can be done using a fixed current injection of 1 or 2 A and the points checked by decreasing the injected voltage.

Most relays have a tolerance of ±5% in impedance reach measurements and ±5% in time measurements. The measured values are checked to ensure they are in the specified range as specified by the manufacturer.

Care should be taken to check the earthing of the CT and Potential Transformer (PT) at the line side or busbar side when injecting the current from the tester in the positive or negative direction.

Residual compensation factor K_0 for earth fault should be input to the relay test set in an old set by the knob setting and in a new advanced test set by loading the relay test file configuration to the tester by a laptop connection.

5. Trip time test at line angle, check t_1, t_2, t_3

   Check the Operating times for each zone, this means Z_1, Z_2, Z_3.

   Fill out Table 20.3.1:

   For each zone 1, zone 2, zone 3 reaches.

   Note: There is an 80 degrees line angle for routine maintenance test.

   The last table will be filled completely for the commissioning test for each zone reach of the relay.

   The same will be done for reverse Z_3 reach, which is checked at point 7 for pH-E and point 8 for pH–pH characteristics. This zone is used in

TABLE 20.3.1 Relay Test Sheet 1

| Angle (degrees) | R–E | Y–E | B–E | R–Y | Y–B | B–R |
	Z	Z	Z	Z	Z	Z
40						
80						
120						
160						
200						
240						
280						
320						
360						

distance protection schemes as a permission overreach blocking scheme to detect reverse faults and this zone is not used to trip the line. Tripping of the line is only in the forward direction.

Table 20.3.2 can then be completed, and the line angle in reverse zone = 80 + 180 = 260 degrees.

TABLE 20.3.2 Relay Test Sheet 2

Angle (Degrees)	Z_3 R	Z_3 Y	Z_3 B	Z_3 R-Y	Z_3 Y-B	Z_3 B-R
260						

The new test set generates an automatic report for the test results and can show every point in the test passed or not passed. The test is correctly passed based on the criteria set by the test set user.

6. Power swing detection

At fault Condition the rate of change in Impedance $\frac{dz}{dt}$ is very high, but at power swing $\frac{dz}{dt}$ is very slow. We can simulate this by decreasing the impedance seen by the relay in slow $\frac{dz}{dt}$ between Z_{ps} and Z_3 Then the relay should block operation of zones 1, 2, and 3, but if we change the rate of $\frac{dz}{dt}$ to be very fast then the relay will not block and will trip by Z_1, Z_2, or Z_3 operation according to the fault location.

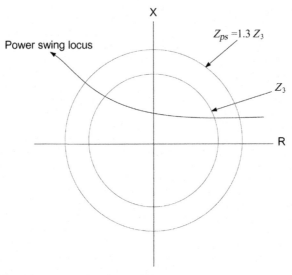

FIGURE 20.3.3 Relay power swing characteristic.

The results in Table 20.3.3 should be verified as per the injected values to the relay.

TABLE 20.3.3 Relay Test Sheet 2

Injected Impedance	Relay Action		Remarks
Z pass between Z_{ps} and Z_3 characteristic very fast	Trip	Yes	Trip based on fault location with Z_1 or Z_2 or Z_3 as injected fault
	Block	No	
Z pass between Z_{ps} characteristic and Z_3 characteristic very slow	Trip	No	Block Z_1 and Z_2 and Z_3 operation after 5 s
	Block	Yes	

7. Switch on to fault instantaneous high-speed OC function test

During this test we will stop operation of the distance function to be switched off and also backup OC function to be switched off.

We simulate this fault as a three-phase fault-close up to the relay location for Example, closing of Transmission Line (TL) on a movable earth cables put on three phases of the TL for maintenance purposes but forgotten on the line after maintenance procedures are finished.

I_{pickup} for Switch on to fault (SOTF) = 0.1 A > 5% I_N

As $I_N = 1$ A

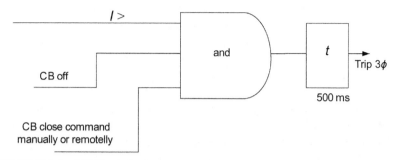

FIGURE 20.3.4   SOTF Logic in distance protection.

Simulate the CB being switched off by binary input from CB auxiliary contact for more than 110 seconds.

A close command is given during this time, then SOTF is enabled for 500 ms. During this time if the measured current is more than 2 A then the SOTF OC relay will operate and bypass the operation of distance relay element. Refer to SOTF Logic in Fig. 20.3.4.

8. Fuse failure blocking function

If a symmetrical zero-sequence voltage $3V_0$ is detected and a symmetrical zero-sequence current $3I_0$ is detected, then the fuse failure function will operate after 5 seconds, during this time if unsymmetrical $3I_0$ is detected the relay will reset the fuse failure blocking of the distance relay as there is a real fault on the system. Some manufacturers also switch on to instantaneous OC relay protection during blocking of distance protection by the fuse failure function.

Also, distance protection is blocked when the Minature Circuit Breaker (MCB) of the VT secondary circuits are tripped due to a fault on the VT secondary. This is provided to the distance protection as binary inputs from MCB's auxiliary contacts.

9. Directional load tests

After energizing the line and the phasing test check is completed then this test is carried out on at least 10%−25% of normal load current of the circuit. The normal load circuits are resistive or resistive-inductive loads. The load will be in phase or lagging of the voltage.

When CT's earthing of the star point is in the line side direction for the above inductive resistive load the direction of distance measurement should be in first quarter as shown in Fig. 20.3.5.

The values of MW, MVAR, voltage, and current phase angle are tested if the direction of the load is as per Fig. 20.3.5. The direction of distance protection measurement is correct, otherwise check that the setting on the distance protection in the line side for CT's star point if the direction is not correct then we should switch off the circuit to modify the CT star point

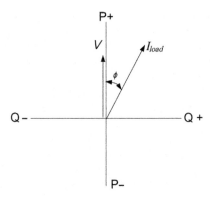

FIGURE 20.3.5   Distance relay directionality test.

connection to the correct direction—normally (Line Side LS) as per the setting in distance protection. Then we repeat the test and measure again to be sure that the direction is correct.

There is also in the market a tester which can do this test when connected to the secondary currents and voltages of the relay.

10. Tele protection scheme test
   a. Tripping scheme.
      Permissive underreach scheme test.
      Permissive overreach scheme test.
      Direct transfer trip test.
   b. Blocking scheme.
      Permissive overreach scheme test.
      Each scheme tripping and logic scheme is simulated by injecting the required zone fault and simulating the received signal.
      Also, a directional earth fault protection scheme will be simulated.
      During a distance protection scheme test the earth fault protection function should be switched off, also during the Directional Earth Fault (DEF) scheme test the distance protection function should be switched off.
11. Autoreclose scheme tests

   A simulation test will be done to check the following:
   Dead time (time between Auto-Reclosure (A/R) start and reclosing shot).
   Reclaim time (time between the first close command and the recloser to be in ready condition).
   External autoreclosure on/off control.
   External A/R blocking.
   A/R blocking for an open CB closing on to a fault.
   A/R blocking for CB in a not-ready condition.
   A/R blocking due to lack of synchronization verification.
   Operation counter check.

Subchapter 20.4

Line Differential Protection Testing and Commissioning

20.4.1 OLD-TYPE PILOT WIRE DIFFERENTIAL PROTECTION TESTING

As shown in Fig. 20.4.1 for simple pilot wire differential protection, this system has a summation CT convert 3ϕ values from CTs to a single-phase quantity to be transferred to the other end by a pilot wire.

1. Secondary insulation resistance test

    By disconnecting the pilot wire from both sides of the relays the insulation between the pilot wire and earth for each wire is measured.
2. Pilot loop resistance tests

    The pilot wires are disconnected at both sides of the line and shorted at one side, then the resistance of the pilot in the other side of the line is measured, including the values of padding resistors and without padding resistors.
3. Pilot connection checks

    Check that the insulation transformer isolates the pilot at the two ends of the line. This transformers are used as isolation transformers and are very important for the safety of persons working on the relay.
4. Secondary Injection Tests

    4.1 Operation and Tripping Current Values Secondary Injection Test

    Open the CT links and trip links and inject the current between the phase and neutral to the relay and increase the injected current until the relay

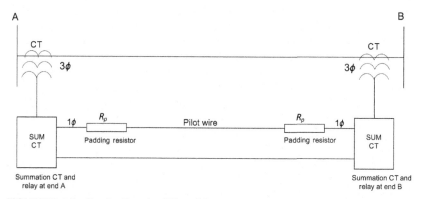

FIGURE 20.4.1   Simple pilot wire differential protection.

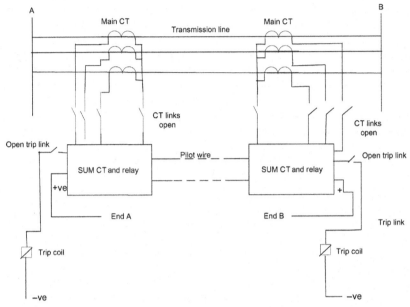

FIGURE 20.4.2   Pilot wire line differential protection for AC and DC circuits.

TABLE 20.4.1 Test Sheet 1

Injected Phase	Relay Operation Current End A	Relay Operation Current End B
R–N		
Y–N		
B–N		
R–Y		
Y–B		
B–R		

operates. Repeat this test for the other two phases, then repeat the test for phase—phase injection and measure the value at which the relay will operate, as shown in Fig. 20.4.2 and Table 20.4.1.

Repeat the above test, but put back the trip links and check the value of the current at which the relay will trip at one phase at least, and check the same at the other end of the line.

5. Load Tests

5.1 Current Transformer and Polarity Tests

This test is done by load current when its value is above 25%−50% of the $I_{nominal}$ of the relay.

Remove the tripping links at both sides of the relay.

Measure the current input to the relay for each phase and in the neutral wire of the CT at both ends of the line, as shown in Fig. 20.4.3.

The reading in each phase should be near 1 A for CT secondary rated 1 A and near 0 in a neutral connection of the CT.

5.2 Stability Load Test

1. SC (Make a short circuit) between the two phase Y and B at both ends of the line to simulate on external fault in phase R of the relay with a load current of 25%−50% I_N check that the relay will remain stable.

2. Check the operation of the relay when short-phase R with neutral at one end or reverse the pilot wires at one end of the line to make the protection unbalanced.

6. Pilot Wire Tests

1. Make SC on a pilot cable. Check the operation of the pilot wire supervision relay and its alarm.

2. Open the pilot wires and check the operation of the supervision relay and its alarm.

3. Reverse the pilot wires at one end only and check the supervision relay will not operate.

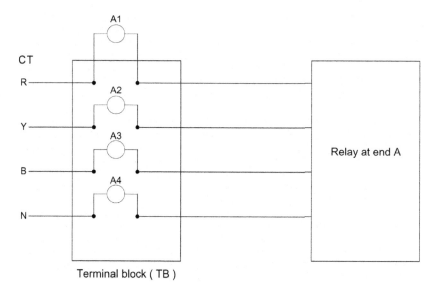

FIGURE 20.4.3 Load test measurements.

4. Note that the supervision relay injects a DC voltage to monitor the pilot cable condition.
5. This test should be done with the HV line switched off.

20.4.2 NEW TYPE LINE DIFFERENTIAL PROTECTION WITH FIBER OPTIC LINK TESTING

1. Secondary injection tests
    1.2 Pickup/drop off test
    This test is done after we set the relay temporary in the loop back test mode as the send optical signal (T_X) is forward to the receive (FO)-Fiber Optic-port R_x to simulate the operation of the commissioning channel and to test the relay alone in one end only refer to Figs. 20.4.4−20.4.5).

FIGURE 20.4.4 Loop back simulation test.

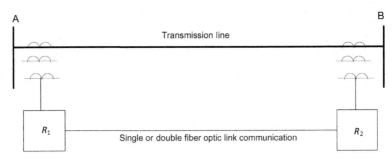

FIGURE 20.4.5 Fiber optic line differential relay.

The loop back mode can also be done by a relay setting tool on Machine-Human-Interface-Display (MHI).

Inject the relay with a current after isolation of the CTs and check the pickup and drop off values for each phase to neutral and for each pH−pH injection, and record the results in a table.

1.3 Current differential bias characteristic
Most modern relays have a characteristic as shown below in Fig. 20.4.6:
$I_1 = 0.2\ I_N\ \ I_2 = 2\ I_N$
$K_1 = 30\%\ \ K_2 = 150\%$
Inject a current in phase R as a bias current and a current in phase Y as a differential current, as shown in Fig. 20.4.7.

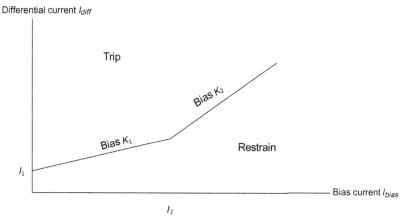

FIGURE 20.4.6 Biased differential protection characteristic for transmission lines.

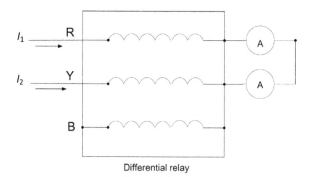

FIGURE 20.4.7 Test of relay bias characteristic.

A drawing of the relay characteristic should follow the relay characteristic provided by the manufacturer.

1.4 Differential relay operating time

Connect the relay and secondary injection tester as shown in Fig. 20.4.8.

Inject current of more than 0.2 A and check the operating time of the relay repeat. This test is used for all phase-N and pH−pH injection. Complete a table with the results.

1.5 On-load test

When the line is energized and loaded, measure the differential current, bias current for each phase on both ends of the line directly on the relay on the MHI. Then reverse the CT connection star point from the line side to the busbar side and check the value of the differential current which should be raised. This test should be done with all tripping links of the relays at both sides of the line opened.

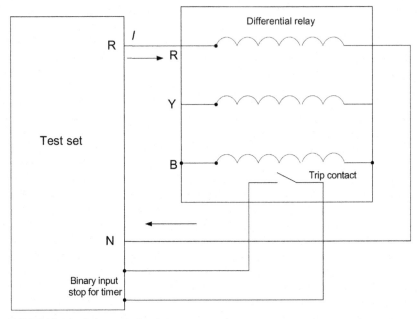

FIGURE 20.4.8 Relay operating time test connection.

Subchapter 20.5

Transformer Protection Testing and Commissioning

20.5.1 TRANSFORMER DIFFERENTIAL PROTECTION

1. Open the tripping links of the relay and, if the circuit is energized, SC the secondary circuits of the CT at both sides of the power transformer and inject current in the relay side only, as shown in Fig. 20.5.1. Consider a power transformer of 500/220 kV, 500 MVA, Yy_0.

1. Close links: A, B, C in TB1.
2. Close links: a, b, c in TB2.
3. Open links: 1, 2, 3, 4 in TB1.
4. Open links: 1, 2, 3, 4 in TB2.

Inject currents are as shown in Fig. 20.5.1 and also connect the output tripping contacts of the relay to the test set as shown in Fig. 20.5.1.

20.5.2 SECONDARY INJECTION TESTS

1. Pickup test

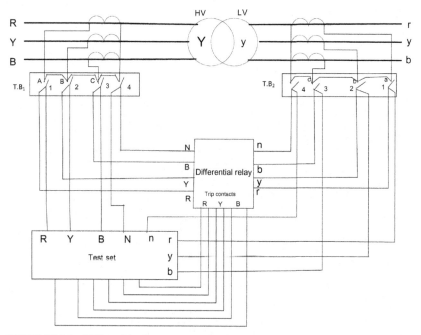

FIGURE 20.5.1   Differential relay test connection.

This test is performed to determine the minimum current required to only operate the relay, the normal limit for pickup value is ± 5% of the nominal pickup value.

a. Inject a current in one side of the transformer CT in different phases, then check the pickup value operation to an output contact on the relay for phases R, Y, B, RY, YB, and BR, and complete a table with this value.

2. Operation characteristic slope test

This test is used to measure the percentage of unbalance that must occur for the percentage slope set on the differential relay to operate.

The test is performed by gradually increasing the amount of unbalanced current applied to the relay. The point at which the relay operates is compared to the relay setting. This is done by injecting a current on both sides of the relay [HV side and low voltage (LV) side] and comparing the measured point for operating and blocking around the relay characteristic to the calculated point from the equation of the relay characteristic given by the relay manufacturer; most transformer differential relays have characteristics as shown in Fig. 20.5.2.

About 12 points should be chosen around the relay characteristic curve to be checked to determine the exact operating point of the relay, and they should have an accuracy normally of about ± 5%.

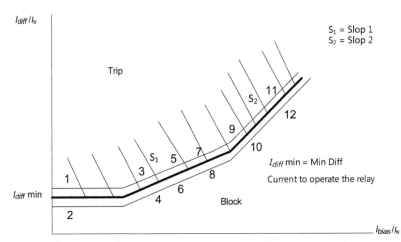

FIGURE 20.5.2 Differential relay characteristic.

TABLE 20.5.1 Test Sheet 1

Injected Current HV Side	Injected Current LV Side	I_{bias} Calculated	I_{bias} Measured	I_{diff} Calculated	I_{diff} Measured	Error %
I_R	I_r					
I_Y	I_Y					
I_B	I_b					

HV, high voltage; LV, low voltage.

Then Table 20.5.1 should be completed and the two values measured and calculated should be compared to check the relay characteristic operation.

3. Trip time test

By injecting the relay with about 1.5 I_{diff}, the trip time for each phase can be measured (injection will be done from the HV side only as this is sufficient). Then Table 20.5.2 is completed with the measured values.

4. Harmonic restrain test

As we know from previous chapters of this book, the differential relay of the transformer will be blocked in two cases as follows:

1. In the case of a high-value inrush current during energization of the transformer this current can be distinguished from the fault current by the second harmonic current as the inrush current is rich in this harmonic but does not exist in the fault current.

TABLE 20.5.2 Test Sheet 1

Injected Phase	Injected Current	Calculated Trip Time	Measured Trip Time	Error %
R	$1.5\ I_{diff}$			
Y				
B				

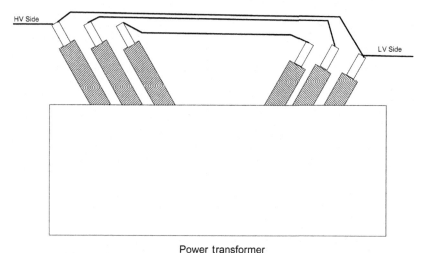

Power transformer

FIGURE 20.5.3 Power transformer with shorting between the HV side and LV side bushings to bypass the transformer impedance. As shown in Figs. 20.5.4 and 20.5.5 an external fault is simulated.

2. When the transformer is overfluxing then it absorbs high current from one side of the transformer, which can lead to a false trip. This current of overexcitation is rich in fifth harmonic, which is not the case for the fault current.

Normally I_{second} harmonic set = 15% I_{diff} and I_{fifth} harmonic set = 30% I_{diff}. Inject the relay with the second harmonic up to 15% Idiff then check that the relay will be blocked above this value, also, check that the relay will blocked above 30% I_{diff} for the fifth harmonic.

20.5.3 PRIMARY INJECTION TEST

In this test we do a stability test by supplying the transformer HV side with 380 V and SC the power transformer at the LV side, if the current is not enough we can SC the power transformer winding between the HV bushing and LV winding, as shown in Figs. 20.5.3 and 20.5.4.

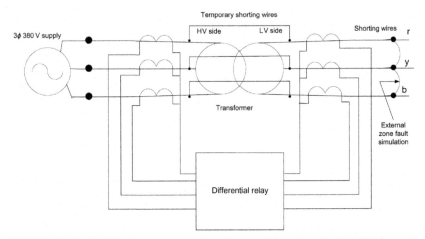

FIGURE 20.5.4 Stability primary test of transformer differential relay by 380 V supply. Simulation of an external fault.

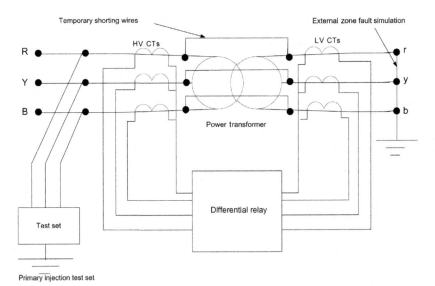

FIGURE 20.5.5 Stability primary test of transformer differential relay by primary test set. Simulation of an external fault.

This test can also be done using a primary inject test set, as shown in Fig. 20.5.5.

For the two methods of injection shown in the two Figs. 20.5.4 and 20.5.5 the current will pass through the CT's of the HV side of the transformer and the LV side CTs of the transformer then we measure the currents in each terminal block in CT secondary circuits in the HV and LV sides of the power transformer differential protection panels, and also the operating

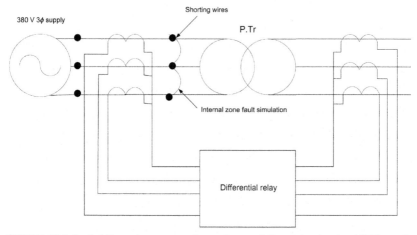

FIGURE 20.5.6 Stability primary test of transformer differential relay by 380 V supply. Simulation for an internal fault.

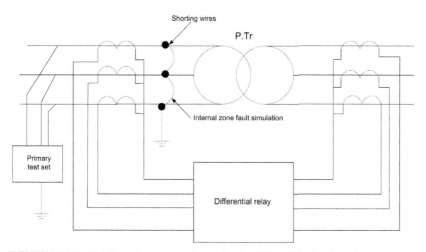

FIGURE 20.5.7 Stability primary test of transformer differential relay by primary test set. Simulation for an internal fault.

differential current in the relay HMI and the bias current on the relay display (HMI) if the differential current is high then the CT star point is not correct and should be corrected at one side of the power transformer as per the design schematic drawing. The test should then be repeated.

For an internal fault we have to simulate an internal fault on the transformer (inside protected zone of the transformer) by placing the shorting wires as shown in Figs. 20.5.6 and 20.5.7.

In this case the differential current should be high and the bias current low, and the relay will trip on this internal zone fault.

Subchapter 20.6

Busbar Protection Testing and Commissioning

20.6.1   SECONDARY INJECTION TEST FOR THE DISCRIMINATING ZONE

The most used scheme is a high-impedance differential protection.

1. Pickup test of busbar protection modules
    Put the busbar protection out of service during this test—this means the buswires will be short-circuited and trip links will be opened.
    Inject the current to each module of the busbar protection to check the pickup value, which should be $\pm 5\%$ of the normal pickup setting of the relay.
2. Pickup test of the breaker failure module if it is attached to the Busbar protection system
    Inject this module with current and check the pickup value is within $\pm 5\%$ of the nominal pickup value. This is done after opening the trip link of the Breaker Failure (BF) system of the substation.
3. Time test measurement of BF relay
    Simulate a BF initiation signal and inject a current above the relay setting, then allow BF time and check the operation of the BF trip—normally it is set to 125 ms. Check that the measured time is $\pm 5\%$ of the time set on the relay.
4. Busbar isolator image test (refer to Fig. 20.6.1).

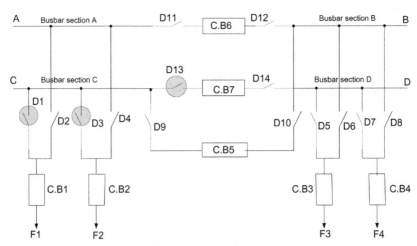

FIGURE 20.6.1   Busbar discrimination test of fault location.

A double busbars with a four-section system has one breaker per feeder.

Check that when D1 for feeder F1 and D3 of feeder F2 are connected to section C, when we simulate a fault on section C with D13 and D14, CB7 are closed and CB1 and CB2 are also closed. This simulates the busbar protection by secondary injection which will trip CB1, CB2, and C.B7. This discrimination of the fault location is determined by busbar protection by the bus isolator image (auxiliary contact from an isolator disconnector, such as D1, which will inform the relay whether this disconnector is connected to bus C or not.

By repeating the last procedures for a fault simulation on sections A–B and C–D with all possible combination of disconnectors status open or close, we can test the discrimination of the busbar protection—this means that the busbar protection will trip only the section which has the fault and leave the other three sections in service.

20.6.2 PRIMARY INJECTION STABILITY TEST

1. Ratio check for a reference circuit

In this test we define a reference circuit as feeder F1, then we check the polarity of CT connections by injecting each CT phase with a primary current, as shown in Fig 20.6.2.

We then measure the reading of I_2 and I_1. I_1/I_2 should correspond to the CT ratio.

Repeat the last test for the ratio for phase Y and phase B.

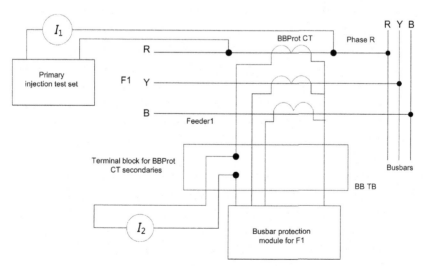

FIGURE 20.6.2 Ratio test for reference circuit F1.

2. Phasing of the reference circuit CT

As in the ratio test we choose F1 (feeder 1) as a reference circuit, then we check the correct polarity of the reference feeder CTs by injecting a current through a primary test set through two phases, as shown in Fig. 20.6.3.

The correct polarity of the reference circuit CT connection exists when the reading of I_2 in the neutral circuit of CTs is near zero, otherwise the polarity will not be correct and we must review the CT connection and star point connection of the CTs. This test is done with temporary SC of R, Y, and B phases, as shown in Fig. 20.6.3.

Repeat the above test between phases Y and B, then between phases B and R. All should read a few milli-Amperes or near zero for the correct polarity of the CT connection on feeder F1.

Now we have a reference circuit F1 checked for ratio and polarity, then we check other circuit CTs by injecting a primary current between the reference circuit and other circuits one by one as F2, as shown in Fig. 20.6.4.

Then, by comparing I_2 and I_4 for a ratio check and I_3 and I_5 for a polarity check, the test between the reference circuit F1 and other feeders F3 and F4 is repeated.

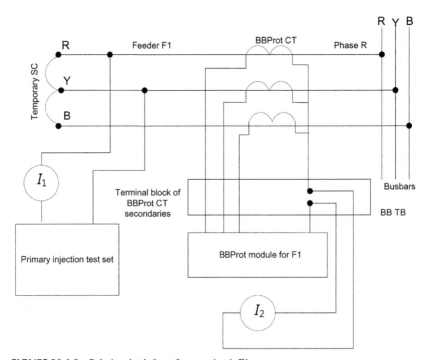

FIGURE 20.6.3 Polarity check for reference circuit F1.

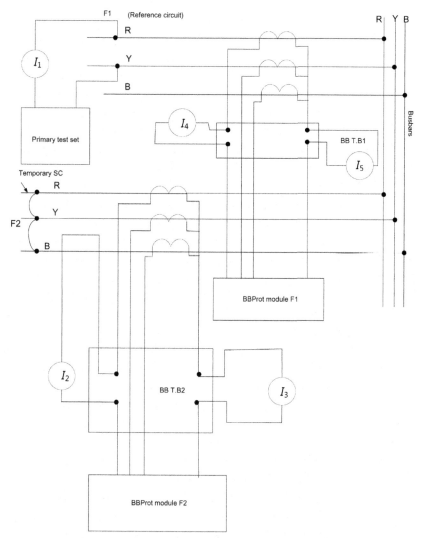

FIGURE 20.6.4   Primary injection of reference circuit feeder F1 and circuit feeder F2.

For a bus-section circuit we make a temporary SC on R, Y, B on any other circuit and check the reference circuit F1 and bus-section circuit with CT on the right side of the bus section as shown in Fig. 20.6.5.

3. Sensitivity test

This test is the same test as for relay module pick up, but it is done by primary injection for each module. For example, for an F1 feeder it can be done during the ratio check test as described in point 1 of Section 20.6.2.

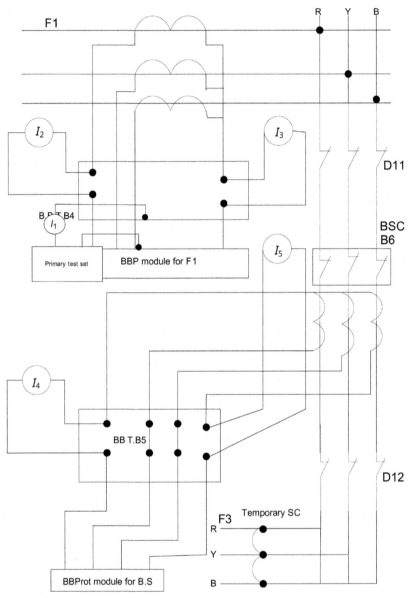

FIGURE 20.6.5 Primary injection of reference circuit F1 and bus section Bus Section (BS).

4. Trip test

This test measures the operating time of BBProt, including trip relays time, and measures the overall operating time to trip the CB practically in the BBProt scheme.

The tests in Sections 20.6.1 and 20.6.2 are repeated for the check zone if a separate CT is provided for this check zone.

5. CT supervision test

This unit is overcurrent relay and is set to monitor the differential circuit for each phase and to give an alarm after a time delay of about 3 seconds when one CT secondary circuit is detected as open and to switch off that zone of the busbar protection. This can be tested by injecting a current greater than the required setting for 3 seconds then checking the operation of the relay to SC the buswires which this CT is connected to and putting this faulty zone of the CT out of service.

Subchapter 20.7

Synchronizing Relay Testing and Commissioning

20.7.1 INTRODUCTION

When there are two parts of a power system network, each has a separate generation to be connected together. This connection should avoid any damage to the power system components or shock loads.

The voltage, phase angle, and frequency difference between the two connected systems should be within the acceptable limits. There are two methods for synchronizing:

1. Check synchronizing

When the two systems are tied by an open breaker then the two frequencies will be different and the phase angle difference will increase (slow slip).

2. System synchronizing

When the two systems are asynchronous, as two generation islands, the rate of slip between the two frequencies will be high. Then the system synchronizing limits are applied. In this case, the phase angle limit will be narrow and closure of the CB will be allowed when the phase angle difference decreases.

Refer to Fig. 20.7.1 to check the synchronizing and system check operation.

Check synchronizing limits:

$\Delta F = 50$ mHz

$\phi = 35 \rightarrow 60$ degrees

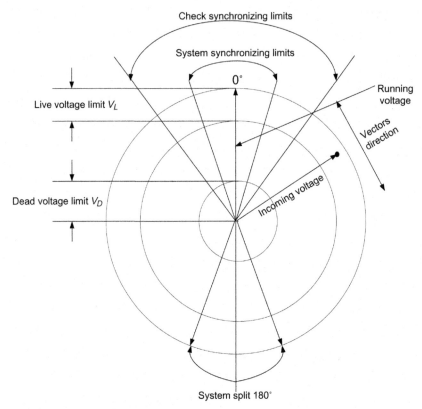

FIGURE 20.7.1 Checking the synchronizing and system check operations.

System synchronizing limits:

$\Delta F = 125$ mHz

$\phi = 10$ degrees

When the two systems are connected outside the permissible limit the following can happen:

Vibration for generating units;

Sudden load for nearby generating stations;

A sudden change in the power system transmission line impedance;

Live voltage: the limit of voltage at which we can consider the line to be on (energized);

Dead voltage: the limit of voltage at which we can consider the line to be switched off.

System synchronizing is sometimes called autosynchronizing.

The old system applied manual synchronizing by synchroscope or lamp method, but the new one applies synchronizing relay for automatic synchronizing operation of a CB between the two electrical parts or systems.

20.7.2   SYNCHRONIZING RELAY TEST

A relay connection is shown in Fig 20.7.2, where:

V_1 = running voltage;

V_2 = incoming voltage.

The settings are:

$\Delta F = 20$ mHZ $\Delta \phi = 30$ degrees

$\Delta V = 10\%$ blocking voltage $= 45\%$ V_n

Live line $= 55\%$ V_n dead line $= 25\%$ V_n

Live bus $= 55\%$ V_n dead bus $= 25\%$ V_n

Connect the circuit as shown in Fig. 20.7.2 then perform the following tests.

1. Differential voltage pickup test

    Inject a fixed voltage of $\frac{110}{\sqrt{3}} = 63.5$ V as the running voltage with a fixed angle of 5 degrees between the two voltages. Then inject the incoming voltage with the relay set for $\Delta V = 5\%$ and check the pickup value for the running voltage and pickup value of the incoming voltage. Repeat the last steps for $\Delta V = 7.5\%$ and $\Delta V = 10\%$ and complete Table 20.7.1.

2. Differential angular pickup test

    Inject the running voltage (bus voltage) and incoming voltage (line voltage) at different phase angles, then check the pickup value of the phase angle as shown in Table 20.7.2. This will be at fixed F, which means $\Delta F = 0$.

3. Differential frequency pickup

    Inject V_1 and V_2 at fixed $\Delta \phi = 10$ degrees, then check the $\Delta F = 20$ mHZ setting and complete Table 20.7.3.

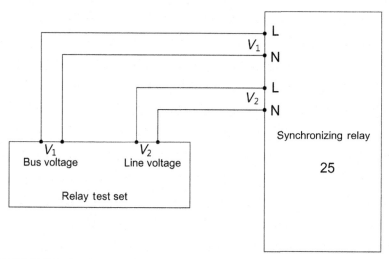

FIGURE 20.7.2 Synchronous relay test connections.

TABLE 20.7.1 Differential Voltage Pickup Test

$\Delta V\%$ Set	ϕ (degrees)	Running Voltage V_1	Incoming Voltage V_2	V_1 Pickup	V_2 Pickup
5%	5	63.5	63.5		
7.5%	5	63.5	63.5		
10%	5	63.5	63.5		

TABLE 20.7.2 Differential Angular Pickup Test

$\Delta\phi$ Set (degrees)	V Line V_2	V Bus V_2	Phase Angle Pickup	
			Lead Side	Lag Side
10	63.5	63.5		
20	63.5	63.5		
30	63.5	63.5		

TABLE 20.7.3 Differential Frequency Pickup Test

ΔF Set	V_1	V_2	$\Delta\phi$ (degrees)	ΔF Pickup
20 mH	63.5	63.5	10	

4. Blocking voltage test

Reduce the bus voltage and line voltage to 45% of $V_n \rightarrow 28.575$ V. Then check that the relay will block operation at this level of voltage and below, and that it will operate above this voltage.

5. Check the live and dead voltage monitor for bus and line voltage

This feature monitors the running and incoming voltage and considers the line live for voltages above 55% V_n and considers the line dead for voltages less than 25 %V_n. The same applies for the bus voltage monitor. This feature allows bypassing of the synchronizing operation for dead conditions of the line or bus.

Subchapter 20.8

Out-of-Step Relay Testing and Commissioning

20.8.1 INTRODUCTION

This relay uses an Ohm relay characteristic to detect the power swing condition and to trip the system and isolate it at certain points (out-of-step tripping scheme) (refer to Fig. 20.8.1).

When impedance passes from zone A to zone B to zone C, this is considered as a power swing and a trip will occur.

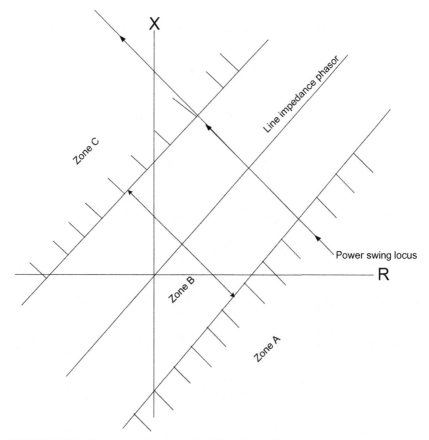

FIGURE 20.8.1   Power swing locus in an R–X impedance diagram.

However, when impedance passes from zone A to zone B only or from zone C to zone B only, this is considered as an external faults and no trip will occur.

20.8.2 RELAY CONNECTION

Refer to the following (Fig. 20.8.2), for out-of-step relay test connections

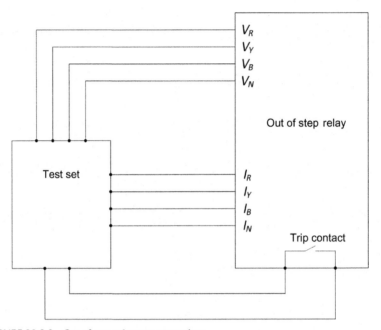

FIGURE 20.8.2 Out-of-step relay test connections.

20.8.3 TEST PROCEDURE

By injecting a fixed current of 1 or 2 A then the value of the injected voltage is changed to simulate the following conditions and the relay response is checked as follows:

Impedance passes from $Z_A \rightarrow Z_B \rightarrow Z_C \rightarrow$ trip;
Impedance passes from $Z_C \rightarrow Z_B \rightarrow Z_A \rightarrow$ trip;
Impedance passes from $Z_A \rightarrow Z_B \rightarrow$ no trip;
Impedance passes from $Z_C \rightarrow Z_B \rightarrow$ no trip.

Chapter 21

A Guided Practical Value of Some Test Results Collected From Actual Power System Testing at Site

21.1 POWER TRANSFORMER TEST

Example 1:
  75/75/25 MVA transformer
  220/70/11.5 kV $F = 50$ Hz
  1. Insulation resistance of winding
  Injected voltage = 5 kV DC
  High voltage (HV)—E   1250 MΩ
  Low voltage (LV)—E   3000 MΩ
  HV—LV   4000 MΩ
  Tertiary—E   3000 MΩ
  HV—tertiary   5000 MΩ
  LV—tertiary   3500 MΩ
  2. Wiring insulation
  Injected voltage = 1000 V
  $R > 1000$ MΩ
  3. Fan motor consumption
  $I_N = 1.8$ Amp
  $I_{Start} = 7.62$ Amp
Example 2:
  125/125/45 MVA
  220/70/11.5 kV
  1. Audio noise level test
  Average level of noise = 77 dB

Practical Power System and Protective Relays Commissioning.
DOI: https://doi.org/10.1016/B978-0-12-816858-5.00021-6
© 2019 Elsevier Inc. All rights reserved.

2. Ratio test (Refer to Table 21.1)

TABLE 21.1 Power transformer turns ratio test results

Tap No.	$\frac{N_1}{N_2}$	Measured Ratios		
9	3.1429	R_N/r_n	Y_N/y_n	B_N/b_n
		3.1429	3.146	3.146

3. No load test (Refer to Table 21.2)

TABLE 21.2 Power transformer no load losses test results

Applied Voltage	Excitation Current	No Load Losses
$V_n = 70$ kV	5.5 A	76 KW

4. Impedance voltage and load losses
LV side SC at 95°C (Refer to Table 21.3)

TABLE 21.3 Power transformer short circuit test results

Applied Voltage		I^2R Losses (kW)	Starry Losses (KW)	Total Losses (KW)	Impedance Voltage (%)
HV	LV				
220	70	350	19	369	17%

5. Winding resistance measurements
$T = 21$°C (Refer to Table 21.4)

TABLE 21.4 Power transformer winding resistance test results

Winding	Terminals	V	I	$R = \frac{V}{I}(\Omega)$
HV	R—Y	5.137	6.280	0.8180
	Y—B	4.196	5.134	0.8175
	B—R	4.192	5.126	0.8178
LV	r—y	0.6814	7.801	0.08735
	y—b	0.6796	7.787	0.08727
	b—r	0.6806	7.794	0.08732

6. Dielectric tests
HV test at power cycle 50 c/s
HV—E → 230 kV
LV—E → 70 kV
Tertiary—E → 28 kV
For 1 minute
7. AC 31.2 kV at tertiary winding at 200 C/S test
For 30 seconds
Tertiary wilding → 31.2 kV at 200 C/S
Then measure HV—E → 460 kV at 200 C/S
Voltage per turn, 2.68 times the rated value.

21.2   HIGH-VOLTAGE TRANSMISSION LINE IMPEDANCE VALUES

220 kV line
  Bundle aluminum
  $R_1 = 0.14$ Ω/km
  $X_1 = 0.4$ Ω/km
  $X_0 = 2.04$ Ω/km

21.3   GAS INSULATED SYSTEM TESTS

Disconnecting switch: contact resistance = 10 μΩ
  Earthing network resistance = 10 μΩ
  Gas insulated system (GIS) main circuit resistance 66 kV between 2 bays with 100 Amp DC injected
  $R = 160–280$ μΩ
  Circuit breaker operating time
  T close = 55 ms
  T open = 33 ms
  Disconnecting switches
  T open < 2 seconds
  T close < 3 seconds
  High-speed earthing switches
  T open < 18 seconds
  T close < 18 seconds
  High-voltage test for GIS
  Example:
  For 132 kV GIS we apply 235 kV for 1 minute.
  For 22 kV SWG we apply 50 kV for 1 minute.

21.4   LOW VOLTAGE PANELS

Torque wrench key = 70 nm.

Chapter 22

Final Substation Primary and Energization and Loading Tests

22.1 FINAL PRIMARY INJECTION TEST OF SUBSTATION

Consider the example of substation A as shown in Fig. 22.1.

This test can be done alone or along with a busbar protection primary test.

By injecting the primary current with, for example, 150 Amps between each of two bays:

1. Between F1 and F2 at point 1 and point 2 with Q_{15} and C.B2 and Q_7 and Q_9 and C.B3 and Q_{16} are closed to test busbar B.B1-A. The test is repeated with Q_{15} and C.B2 and Q_8 and Q_{10} and C.B3 and Q_{16} are closed to test B.B2-A.
2. Between F2 and F3 at point 2 and point 4 with C.B3 and Q_{16} and Q_9 and Q_1 and Q_5 and C.B1 are closed and the test is repeated with an injection between point 2 and point 3 with C.B3 and Q_{16} and Q_{10} and Q_3 and Q_6 and C.B1 closed.

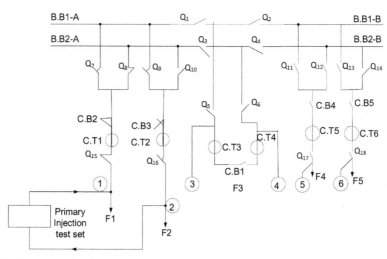

FIGURE 22.1 Substation A.

Practical Power System and Protective Relays Commissioning.
DOI: https://doi.org/10.1016/B978-0-12-816858-5.00022-8

365

3. Between F3 and F4 at point 3 and point 5 with C.B4 and Q_{17} and Q_{11} and Q_2 and Q_5 are closed and the above test is repeated with an injection between point 5 and point 4 with C.B4 and Q_{17} and Q_{12} and Q_4 and Q_6 closed.
4. Between F6 and F5 at point 5 and point 6 with Q_{17} and C.B4 and Q_{11} and Q_{13} and C.B5 and Q_{18} are closed and the test is repeated with Q_{17} and C.B4 and Q_{12} and Q_{14} and C.B5 and Q_{18} closed.
5. For all the above tests the secondary currents in protection, metering, and busbar protection panels for each phase (R, Y, B) are measured and the results recorded in a proper test sheet. All current transformer (CT) phases should be read to confirm the correct values in all panels in the substation based on CT ratios.

22.2 ENERGIZATION OF SUBSTATION TEST

Before energization the following points should be checked:

All commissioning test are completed and have acceptable results;
All shorting links on CTs are removed and all miniature circuit breakers (MCBs) are on;
All relays are set to the final setting;
All trip tests for all relays are done;
No earthing is left on any primary circuit;
Visual checks for power transformers for power cable termination, cable box cover oil level, and cooler fan setting;
Visual check for SF_6 pressure in gas insulated system;
All work permit, switching program; and commissioning overcurrent temporary settings are all ready at the site;
All safety signs are there before energization.

After energization the following points should be carried out:

Secondary voltage measurements to be checked in all points in all panels;
Phase sequence to be checked;
Phasing checked before paralleling two circuits;
After loading the circuit all secondary currents to be measured in all panels and all phases;
Directional test to be done for directional relays such as directional over current (O/C) or Earth fault (EF) relays and distance relays;
Stability test for differential relays during circuit loading by reversing the polarity of the secondary currents and the performance of the differential relay checked;
Load test of automatic voltage regulator for transformer.

22.2.1   Synchronizing/Phasing Checks

Before paralleling of two circuits from the same generating station, the voltage difference between the phases of the two circuits should be approximately zero, as follows:

R_1 to $R_1 = 0$ B_1 to $B_2 = 0$ Y_1 to $Y_2 = 0$

These measurements are done in the secondary voltage of the voltage transformer of the two circuits.

Before paralleling of two circuits from two different generating stations, in addition to the phasing process as above, synchronization must also be done as follows:

Check the correct setting on the synchro check relay;

Check the phasing voltage difference as above between the phases of the two circuits;

Check the synchro check relay will pickup when the synchroscope is in the 12 o' clock position;

Energize the first circuit to an unenergized busbar (no circuits are connected on that busbar) then, before connecting the second circuit to the same busbar, check that the synchroscope pointer rotates (this is rotation of incoming voltage vector) due to the slip frequency between the two generating stations;

When the rotation of the synchroscope pointer is slow and the phase angle between the incoming and running voltage becomes within the permissible range, the synchrocheck relay will close the circuit breaker of the incoming circuit.

22.2.2   Directional Test

This test will verify that the directional protection, such as distance relay, is looking in the line direction only. This is done by a load test as we simulate the load current and voltage in the trip direction and check the relay response, then by reversing the secondary current polarity, and finally checking that the relay will not trip in the reverse direction.

The following points should be taken care of before and after the test:

Before doing the test isolate the tripping contacts of the relay;

After the test ends restore the correct polarity of current transformer (CT) connections.

22.2.3   On-Load Stability Test

1. Transformer differential protection

Read the differential current on the relay screen on the digital relay and also measure it inside the relay wires in old electromechanical relays. This differential current should be near zero by reversing the current polarity on one side of the power transformer CTs. The secondary relay will pickup as the differential current is increased, with the relay before the test isolated from the tripping circuit, then the current polarity of the CTs is restored.

2. Busbar differential protection

With the trip contacts of busbar protection are isolated we can measure the differential current on the relay by shorting the CT input of each feeder connected to the specified zone of protection. We should have a rise in differential current and the relay should issue a trip signal to the bus zone which this feeder is connected to.

Chapter 23

Substation Testing and Commissioning Time Schedules and Resources Management

23.1 INTRODUCTION

In order to complete the substation commissioning work in efficiently and effectively we should manage the activities involved in these tests and manage resources to match each activity with the correct number of people with the appropriate skills and also manage the time to do these activities wisely, as increased time spent inevitably leads to increased costs.

23.2 PRACTICAL EXAMPLE OF EXECUTED SUBSTATIONS

23.2.1 Gas Insulated System Substation

The substation has the following data:

500/220 kV substation

3 bays—500 kV feeders + 2 spare feeders

12 bays—220 kV feeders + 2 spare feeders

3 bays—transformers 500/220 kV and 3 transformers 220/11 kV

20 feeders—11 kV feeders

Refer to the following schedule for activities, required manpower, and required time to completion.

A = no. of feeder bays, B = no. of transformer bays, C = no. of 11 kV feeders.

Note that some activities can be done in parallel (Table 23.1).

$A = 3 + 2 + 12 + 2 = 19$

$B = 3 + 3 = 6$

$C = 11$ kV feeders $= 20$

For a substation in this example (Table 23.2), where A, B, C, D, F, G, P, N, K, and M are the 10 different engineers assigned to do this job.

Total required months = 6 months

Practical Power System and Protective Relays Commissioning.
DOI: https://doi.org/10.1016/B978-0-12-816858-5.00023-X

TABLE 23.1 Substation Example for time and resources management

Number	Activity	Engineer	Technician	Labor	Duration Days	Required Days
1	General inspection	1			1	0
2	Earth resistance measurement		1/A		1	1
3	Power services and control cables, cable trays continuity, and insulation test	1/A	1/A		1	1
4	Batteries and battery charges		1	1	$\frac{1}{2}$	
4.1	Battery discharge and charge test			1	$\frac{1}{2}$	1
4.2	Battery charger test	1		1	$\frac{1}{2}$	
5	Relays and control panels 500 and 220 kV SWG					33
5.1	Line differential protection	1	1		$A*1$	
5.2	Directional O/C and EF relays and check of operation of angle of directional element	1	1		1	
5.3	O/C and EF relays	1	1		1	
5.4	Breaker failure protection	1	1		1	
5.5	Transformer O/C and EF relays	1	1		$B*1$	
5.6	Transformer differential port and REF relays	1	1		$B*1$	
5.7	Automatic voltage regulator test	1	1		1	

No.	Description				
5.8	Transformer EF relay	1	1		1
5.9	Busbar protection secondary and primary tests	1	1		2
5.10	Secondary injection test for meters ammeters, voltmeters PF meters, MW, MVAR meters, and transducers	1	1		1
5.11	Relays and control 11 kV SWG				15
5.12	Directional OC and EF relays and check angle of operation of directional element	1	1	1	$\dfrac{C}{5}$
5.13	OC and EF relays	1	1	1	$\dfrac{C}{5}$
5.14	Transformer differential and O/C and EF relays	1	1		3
5.15	Scheme and trip test	1	1		$\dfrac{C}{5}$
6	GIS test 500 kV				30
6.1	Resistance measurement of the main circuit		1	1	$\dfrac{A+B}{2}$
6.2	Time test of CB operation	1	1		$\dfrac{A+B}{3}$
6.3	Operation test of isolator, earthing switch, and auxiliary contacts	1	1		$\dfrac{A+B}{4}$
6.4	Test of air pressure switch of CB or CB spring		1	1	$\dfrac{A+B}{5}$
6.5	Test of SF_6 gas pressure switch for alarm and trip		1	1	$\dfrac{A+B}{5}$

(Continued)

TABLE 23.1 (Continued)

Number	Activity	Engineer	Technician	Labor	Duration Days	Required Days
6.6	Moisture content check of SF$_6$ gas		1	1	$\dfrac{A+B}{5}$	
6.7	SF$_6$ gas leakage test for GIS	1	1	1	$\dfrac{A+B}{5}$	
6.8	Air or oil leakage for circuit breaker mechanism		1	1	$\dfrac{A+B}{5}$	
6.9	Inter trip of low gas pressure of GIS	1			$\dfrac{A+B}{5}$	
6.10	Test of air compressor or hydraulic system or spring charging and discharging of CB/S	1	1	1	$\dfrac{A+B}{5}$	
6.11	Low air or oil pressure block test		1		$\dfrac{A+B}{2}$	
6.12	Measurement of insulation resistance of CT and VT circuit and confirm of earthing point		1	1	$\dfrac{A+B}{4}$	
6.13	DC winding resistance measurement of CTs		1	1	$\dfrac{A+B}{3}$	
6.14	Kick test for VTs		1	1		
6.15	Excitation characteristic of CT curve		1	1	$\dfrac{A+B}{4}$	
6.16	High-voltage test and VT's ratio check		1	1		

No.	Item				Rate	Amount
7	Test of 220 kV GIS same as above in (6)					30
8	Test of 11 kV switch gear					20
8.1	CB insulation and contact resistance		1	1	$\dfrac{C}{5}$	
8.2	Meaurement of insulation resistance of CTs and VT circuits and confirmation of earthing point		1	1	$\dfrac{C}{10}$	
8.3	DC winding resistance measurement of CTs		1	1	$\dfrac{C}{5}$	
8.4	Kick test of CTs		1	1	$\dfrac{C}{10}$	
8.5	Excitation characteristic of CT curve		1	1	$\dfrac{C}{5}$	
8.6	Kick test of VTs		1	1	$\dfrac{C}{5}$	
8.7	High-voltage test and VT ratio check		1	1	$\dfrac{C}{20}$	
8.8	CB scheme check	1		1	$\dfrac{C}{3}$	
9	Operation indication and interlocking test	1		1		20
9.1	System operation and indication check for 500 kV SWG	1	1	1	$\dfrac{A}{5}$	
9.2	System operation and indication check for 220 kV SWG	1	1	1	$\dfrac{A}{5}$	
9.3	System operation and indication check for 220 kV SWG	1	1	1	$\dfrac{C}{10}$	

(Continued)

TABLE 23.1 (Continued)

Number	Activity	Engineer	Technician	Labor	Duration Days	Required Days
9.4	Tap changer operation check	1	1	1	1	
9.5	Arc protection test of 11 kV SWG		1	1	1	
9.6	500 kV alarm and annunciator check		1	1	$\frac{A}{10}$	
9.7	220 kV alarm and annunciator check		1	1	$\frac{A}{10}$	
9.8	11 kV alarm and annunciator check		1	1	$\frac{C}{10}$	
9.9	500 kV interlock test		1	1	$\frac{A}{5}$	
9.10	220 kV interlock test		1	1	$\frac{A}{5}$	
9.11	11 kV interlock test		1	1	$\frac{C}{5}$	
10	Primary injection test of S/S					10
10.1	CT ratio, secondary circuits check of all protection and metering 500 and 220 kV SWG	1	1	1	10	
11	500/220 kV transformer insulation check					24
11.1	Power transformer insulation check		1	1	$\frac{B}{3}$	
11.2	Cooling fan motor test		1	1	$\frac{B}{3}$	

11.3	Tap changer operation check	1	1	1	1
11.4	Check oil and winding temperature indicators		1	1	$\frac{B}{3}$
11.5	Insulation resistance			1	$\frac{B}{3}$
11.6	Ratio check	1		1	$\frac{B}{3}$
11.7	Winding resistance measurement			1	$\frac{B}{3}$
11.8	Vector group test	1		1	$\frac{B}{3}$
11.9	Dielectric breakdown test for transformer's oil			1	$\frac{B}{3}$
11.10	Transformer oil filtering	1		1	B
12	220/11kV Transformer test same as (11)	24			
13	Auxiliary station transformer (500kVA)	3			
13.1	Insulation resistance measurement			1	$\frac{1}{2}$
13.2	Ratio check			1	$\frac{1}{2}$
13.3	Vector group test	1		1	$\frac{1}{2}$
13.4	Winding resistance			1	$\frac{1}{2}$

(Continued)

TABLE 23.1 (Continued)

Number	Activity	Engineer	Technician	Labor	Duration Days	Required Days
13.5	Winding temperature		1	1	$\frac{1}{2}$	
14	Transformers diff. stability test					2
14.1	500/220kV transformer diff. relay stability test	1	1	1	1	
14.2	220/11kV transformer diff. relay stability test	1	1	1	1	1
15	Line diff. relay stability test	1	1	1	1	
16	Busbar protection primary stability test					20
16.1	Busbar protection primary stability test 500kV SWG	1	1	2	10	
16.2	Busbar protection primary stability test 220kV SWG	1	1	2	10	
17	Low voltage A.C & D.C boards					4
17.1	Operation & indication		1	1	1	
17.2	Under voltage relay test		1	1	1	
17.3	Earth fault relay test		1	1	1	
17.4	Ammeter, voltmeter&C.T test		1	1	1	
18	Scada system test	1	1	1	10	10
19	Telecommunication equipments test	1	1	1		15
20	Fire fighting system commissioning	1	1	4		15

GIS, gas insulated system; CB, circuit breaker; CT, current transformer; VT, voltage transformer.

TABLE 23.2 Summary of substation time and resources management

Number	Activity	Required Engineers	Required Days	Months 1	2	3	4	5	6
1	General inspection	A	1	X					
2	S/S earth resistance measurements	A	1	X					
3	Power services and control cables, cable trays continuity and insulation test	A	1	X					
4	Batteries and battery charges	B	1	X					
5	• 500 kV and 220 kV relays and control panels	C	33				X	X	
	• 11 kV relays and control panels		15						
6	GIS 500 kV tests	D	30	X					
7	GIS 200 kV	D	30		X				
8	11 kV SWG	F	20	X					
9	Operation indication and interlocking 500, 220, 11 kV	C	20		X	X			
10	Primary injection tests of S/S 500 and 220 kV SWG	C	10						X
11	500/220 kV transformer stability test	G	24		X				
12	220/220 kV transformer stability test	G	24			X			
13	(500 kVA) auxiliary transformer test station	G	3		X				

(Continued)

TABLE 23.2 (Continued)

Number	Activity	Required Engineers	Required Days	Months					
				1	2	3	4	5	6
14	Transformer differential stability test 500/220 kV transformer 220/11 kV transformer	C	2						X
15	Line differential relay stability test	F	1						X
16	Busbar protection primary stability test 500 and 220 kV	P	20						X
17	Low-voltage AC and DC board tests	P	4		X				
18	Scada system test	N	10				X		
19	Telecommunication equipment tests	K	15				X		
20	Fire-fighting system commissioning	M	15					X	
21	Substation energization	C	3						X

Ten engineers, fifteen technicians, and fifteen laborers are required to commission this 500/220/11 kV gas insulated system (GIS) substation with:

5 (500 kV bays)	14 (220 kV bays)	20 (11 kV feeders)
3 (transformers 500/220 kV)	3 (transformers 220/11 kV)	

The above data are very complicated and need to be carefully managed in regard to resources and parallel activities, and coordination with other parties to the project, such as the project manager, site installation manager, subcontractors, and the national electricity control center.

Index

Note: Page numbers followed by "*f*" and "*t*" refer to figures and tables, respectively.

Printed in the United States
By Bookmasters